ALTERNATING CURRENTS OF ELECTRICITY

AND THE

THEORY OF TRANSFORMERS

ALTERNATING CURRENTS

OF ELECTRICITY

AND

THE THEORY OF TRANSFORMERS

BY

ALFRED STILL

ASSOC. M. INST. C.E.

WITH NUMEROUS DIAGRAMS

WHITTAKER & CO.

2 WHITE HART STREET, PATERNOSTER SQUARE, LONDON

AND 66 FIFTH AVENUE, NEW YORK

1898

PREFACE

———•◇•———

ALTHOUGH the literature of alternating currents has been considerably added to of late years, the author believes that there is still room for a small book—such as the present one—in which the principles determining the behaviour of single phase alternating currents under various conditions are considered less from the scientist's point of view, and more from an engineering standpoint, than is usually the case.

The book has been written, not only for engineering students, but also for those engineers who are but slightly acquainted with alternating current problems; or who, though their practical knowledge of the subject may be extensive, are yet anxious to get an elementary but sufficiently accurate idea of the leading principles involved, which will enable them to solve many—if not the greater number—of the problems likely to arise in practice.

On account of the unsuitability of analytical methods for the solution of alternating current problems, graphical methods have been used throughout; and the introduction of mathematics has been entirely avoided.

The thanks of the author are due to the ' Electrician ' Printing and Publishing Company, who have kindly allowed him to use some of the blocks illustrating his articles on ' Principles of Transformer Design ' which appeared in the ' Electrician ' in November 1894.

BOWGREEN FARM, BOWDON, CHESHIRE:
            March 1898.

CONTENTS

———◦◦◦———

viii

ALTERNATING CURRENTS OF ELECTRICITY

AND THE

THEORY OF TRANSFORMERS

MAGNETIC PRINCIPLES

1. **Measurement of Magnetic Field.**—Since the magnetic field due to an electric current is of far more practical importance in the case of an alternating current than in that of a continuous current—which, so long as it does not change in strength, produces no changes in the magnetic condition of the surrounding medium—it will be advisable to briefly consider the more important properties of the magnetic circuit. Another reason for introducing here a short review of our present knowledge on this subject is that much want of clearness, and some misconception, prevails in the minds of many as to the more modern ways of regarding magnetic phenomena. The reason of this may be sought in the fact that it is

B

1

customary to consider magnetic effects apart from the
electric currents to which they are due ; and also that
old-fashioned and intricate ways of treating the subject
are constantly to be met with, even in comparatively
modern text-books. ' It must not, however, be supposed
that what follows is intended to take the place of such
text-books, the present object being mainly to arrive at
some clear understanding regarding (1) the relation
existing between the magnetic condition and the electric currents producing it, and (2) the manner in which
this magnetic condition can, in its turn, give rise to
electric currents.

There are many ways in which the existence of a
magnetic field may be detected, and its intensity and
direction measured ; but the property of magnetism
with which we are principally concerned is that in
virtue of which momentary currents are induced in
electric conductors when the strength of the magnetic
field, or its direction relatively to the position of the
electric circuit, is altered.

Consider a flat coil of S turns of wire, enclosing an
area A and having a resistance R.   When such a coil is
suddenly placed in, or withdrawn from, a magnetic field,
a quantity of electricity, Q, is set in motion, which, for a
given position of the coil, and a given constant magnetic condition, is found to be proportional to the
number of turns in the coil and its area A, and inversely
as the resistance R of the circuit.   We may therefore

say that the expression QR/AS is a measure of the magnetic condition, and if we consider the plane of the coil to be normal to the direction of the magnetic field, we may put :

$$\mathsf{H} = \frac{QR}{AS} \qquad \cdot \qquad \cdot \qquad \cdot \qquad (1),$$

where H stands for the *intensity* of the field, or its strength per unit area of cross-section. H has also been called the *induction*, on account of the part it plays in the effect we have just been considering.

For all practical purposes we may take it that the magnetic condition due to a given strength of current flowing in a given arrangement of conductors is the same, whatever may be the material surrounding these conductors, with the exception only of iron, nickel, and cobalt.

When the magnetic condition is measured in iron, the induction is no longer denoted by H but by B; we shall, however, consider the reason of this later on.

2. **Lines of Force.**—Faraday was the first to suggest the use of imaginary lines drawn in space to represent (by their direction) the direction of the induction in a magnetic field, and (by their distance apart) the strength or intensity of the field. He called these lines *lines of force*, an expression which is sometimes objected to on the ground that it is inaccurate, though it is difficult to see wherein the inaccuracy lies ; but it is undoubtedly confusing at times, as the

density of these lines of force is sometimes proportional
to the magnetising force, i.e. the ampere-turns producing the field, whereas at other times—when there
is iron in the magnetic circuit—this proportionality
ceases to exist. We shall, therefore, in accordance
with present-day practice, avoid the expression ' lines
of force,' and speak only of *lines of induction*, or, more
simply, after Dr. S. P. Thompson, of magnetic lines.

What is called a magnetic line must always be
thought of as being the centre line of a unit *tube* of
induction, the characteristic feature of a tube of induction being the constancy of the magnetic flux through
it. It follows that, if we adopt the system of absolute
C.G.S. units, a field of unit strength would be represented by one line to every square centimetre of cross-
section. Since the direction of the lines of induction is
the same at any point as the direction of the resultant
magnetic induction at that point, it follows that, although these lines may be conceived as being very close
together (in strong fields), they can never cross each
other, because the resultant induction at any point cannot have more than one definite direction. Also, it is of
fundamental importance to remember that every line of
induction is always closed upon itself, hence we may
speak of the *magnetic circuit* as we do of the electric circuit, and the two are in many ways analogous.

From what has been said it follows that, in any
magnetic circuit, the total flux of induction is the same

through all cross-sections of such a circuit. This principle, which is called the *principle of conservation of the flux of induction*, is mathematically defined when we say that the surface integral of the flux of induction over any cross-section of a tube of induction is always constant. In case any wrong meaning should be attributed to the terms *magnetic flux* or *total flux of induction*, it may be as well to state that these expressions are derived from Lord Kelvin's analogies between magnetism and the flow of liquids through porous bodies.

We are now in a position to again consider the meaning of equation (1). It is evident that, in order that H may be expressed in absolute C.G.S. electromagnetic units (which is the usual practice), Q, R, and A must be expressed in the same units. Therefore, if Q is measured in coulombs, R in ohms, and A in square centimetres, it follows that

$$H = \frac{QR \times 10^8}{AS}.$$

If now we substitute for Q its value $i \times t$, where t stands for the time (in seconds) taken to enclose or withdraw the magnetism, and i is the average current (in amperes) flowing through the circuit during the time t; and, further, if we put in the place of $i \times R$ its equivalent E, which represents the average value of the E.M.F. generated in the coil by the introduction or

withdrawal of the magnetism, we finally obtain the
expression

$$\frac{\mathsf{H}A}{t \times 10^8} = \frac{E}{S} \qquad . \qquad . \qquad . \qquad .(2),$$

by which we see that the E.M.F. generated in the
conductor is proportional to the rate of cutting (or of
enclosing or withdrawing) of magnetic lines ; and if we
imagine the coil to consist of a single turn of wire, i.e.
$S = 1$, the above equation becomes simply another way
of stating the well known rule that one hundred
million C.G.S. lines cut per second generate one volt.

3. Magnetic Fields due to Electric Currents.
Of the many means at our disposal for measuring the
strength of a magnetic field, the method just described—
in connection with which a ballistic galvanometer is used—
is sufficient to show that no difficulty need be experienced
in determining this quantity, which has been called H.
We are therefore in a position to consider the relations
existing between the magnetic field and the electric
currents which produce it.

In the case of a long straight wire carrying a current,
the intensity of the magnetic field in the neighbourhood
of the wire varies directly as the strength of the current,
and inversely as the distance from the conductor of the
point considered. This relation was first experimentally
proved by Biot and Savart, and sometimes goes by the
name of Biot and Savart's law. The exact expression
is :

$$\mathsf{H} = \frac{2i_a}{r}$$

where i_a is the current in absolute C.G.S. units, i.e. in deca-amperes ; r being, of course, in centimetres.

Imagine the conductor to be now bent into the form of a complete ring : the expression for the strength of field at the centre—that is to say, at a point distant r centimetres from all parts of the conductor—is

$$H = \frac{(2\pi r)i_a}{r^2}$$

by which the field at the centre of a tangent galvanometer may be calculated, and which naturally leads to the definition of the absolute unit of current ; this being defined as a current of such strength that, when 1 cm. length of its circuit is bent into an arc of 1 cm. radius, it creates a field of unit intensity at the centre of the arc.

But the case of the greatest interest to us at present is that of a long solenoid, uniformly wound with S turns of wire per centimetre of its length. The field in the middle portions of such a solenoid is uniform—that is to say, it has the same strength and direction at all points within the coil which are not taken too near the ends, and it is expressed by the formula :

$$H = \frac{4\pi}{10} Si \qquad . \qquad . \qquad . \quad (3),$$

where $Si =$ the ampere-turns *per centimetre.*

Towards the ends of the solenoid the intensity

of the field falls off rapidly, until, at the extreme ends, it is only equal to half the above value.

The expression (3) is equally true of a closed coil ; by which is meant a solenoid bent round upon itself until the starting and the finishing ends meet. The similarity between such a coil and a straight solenoid of infinite length will be better understood after we have considered the question of the resistance of a magnetic circuit.

4. **Electrical Analogy — Magneto-motive-Force.**—Consider a solenoid bent round in the manner above described so as to take the shape of what is generally known as an anchor ring. Let its mean length be l, and suppose that it is uniformly wound with S turns per centimetre. Then by formula (3)

$$\mathsf{H} = \frac{4\pi}{10} Si,$$

and if the cross-section is A sq. cms., the total flux of induction, which we will denote by N, will be $\mathsf{H} A$; hence $\mathsf{H} = \dfrac{\mathsf{N}}{A}.$

Substituting this value for H, and multiplying both sides of the equation by l, the length of the coil, we finally obtain the expression :

$$\mathsf{N} = \frac{\frac{4\pi}{10} Si \times l}{\frac{l}{A}} \quad . \qquad . \qquad . \qquad . \ (4).$$

In order to clearly understand the meaning of this equation, let us suppose the magnetic arrangement which we have been considering to be replaced by a glass tube filled with mercury, through which a steady current of i amperes is flowing. By Ohm's law:

$$i = \frac{E}{\frac{l}{A} \times \rho}$$ where $E =$ the E.M.F. producing the flow of

current and $\rho =$ the specific resistance of the mercury. This law we put into words by saying that in an electric circuit the total resulting flow of current is obtained by dividing the resultant E.M.F. in the circuit by the electrical resistance of the circuit. So also, in equation (4), if we denote the quantity $\frac{4\pi}{10} Si \times l$ by M, and call it the *magneto-motive force*, we may likewise say that in a magnetic circuit the total resulting flux of magnetism is obtained by dividing the resultant M.M.F. in the circuit by the magnetic resistance of the circuit. The specific magnetic resistance of the material filling the interior of the coil, or its reciprocal, the *permeability*, μ (a term which, again, owes its origin to Lord Kelvin's hydrokinetic analogy), does not appear in equation (4) because in this case the space inside the winding is supposed to be occupied by a ' non-magnetic ' material such as air, of which—on account of our choice of units—the permeability must be taken as unity.

It is, perhaps, needless to remark that this analogy

between the magnetic and the electric circuits must not be carried too far; but it is exceedingly useful, especially to the engineer, who generally has to determine the total flux N in some particular portion of a magnetic circuit.

It should be mentioned that the term 'magnetic resistance,' which we have used on account of this analogy, is sometimes objected to on the grounds (a) that it is usual to associate ideas of loss of energy with the term 'resistance,' whereas the maintenance of a magnetic field, whatever may be the resistance of the magnetic circuit, does not involve expenditure of energy; and (b) that the magnetic resistance of iron, nickel, or cobalt does not depend solely on the nature of the material, but is also a function of the induction, whereas in the electrical analogy the resistance is in every case an attribute of the material itself, and has a physical meaning apart from its definition as the ratio of E.M.F. to current.

Having stated these objections, and thereby removed any possibility of misconception on the part of the reader, we shall proceed to consider the effect of replacing the non-magnetic core in the arrangement under discussion by an iron one.

5. **Magnetic Induction in Iron.**—Assuming the magnetising force to be the same as before, the induction is now greater than it was when the interior of the helix was occupied by air, and it is

denoted by B. The ratio B/H gives us the permeability, or the *multiplying power* of the iron. This, as is well known, is not constant, even for a given sample of iron, but depends upon many things, among others on the magnitude of the magnetising force and the previous magnetic condition of the iron. The relation of B to H, or of B to the exciting ampere-turns, must therefore always be determined experimentally; it is usually expressed by the aid of a curve, the general characteristics of which are too well known to necessitate their being dwelt upon here.

In the expression $B = \mu H$ it is customary to define H as the number of lines per square centimetre which the magnetising coil would produce in the space occupied by the iron, on the assumption that the iron core were removed, the resultant magnetising force remaining the same as before. The point which is not generally clearly explained is that there is no necessity whatever to consider the iron core removed, or even to imagine longitudinal holes drilled through the mass of the iron, in order to understand what is meant by H in the above relation. There is no such thing in nature as an insulator of magnetism, and the magnetic intensity represented by H is a function only of the resultant magnetising force, or difference of magnetic potential, and the geometrical dimensions of the magnetic circuit, or portion of magnetic circuit, considered; it has just as real an existence in a mass of iron which has practically reached its saturation limit as in a vacuum, or in air ; and when we say that the number of lines represented by B are made up of the number of lines represented by H *in addition* to the number of lines due to the magnetic condition of the iron, this is not an arbitrary or artificial division, but, on the contrary, a scientific analysis which shows what actually takes place. It is well known that, with very strong magnetising forces, the magnetic condition of iron may be brought very near to the saturation point; but, on the other hand, the quantity represented by B does not approach a saturation value, but increases, to all appearances, without limit.   This is because the component H of the resultant induction B increases always in proportion to the magnetising force.   It is therefore usual to write :

$$B = 4\pi I + H,$$

where the quantity $4\pi I$ represents the number of magnetic lines per square centimetre *added* by the iron.   I is called the intensity of the magnetisation of the iron, and is a measure of that physical condition of the iron to which the additional number of lines $4\pi I$ are due.   The factor 4π is simply a multiplier which depends upon our system of units, the meaning of which does not concern us at present, but is easily understood when dealing with the fields surrounding the ends of bar magnets.

    In practice, especially in connection with alternating current work, the inductions used are generally low, in which case the component H of the induction B may be neglected, as it is very small in comparison with the term $4\pi I$; it is, however, usual to read the values of the induction directly off a curve which gives the relation between B and the resultant magnetising force for the particular kind of iron to which the calculations apply.

6. **The Magnetic Circuit. Straight Bar.**—
In the case, just considered, of a closed iron ring, all the magnetic lines pass through the iron, and, in order that N in equation (4) may still stand for the total flux of induction, we must now write

$$N = \frac{\frac{4\pi}{10} Si \times l}{\frac{l}{A \times \mu}} . \quad . \quad . \quad . \quad (5),$$

where the quantity $l/A\mu$ is still called the magnetic resistance of the circuit. It is hardly necessary to point out that the above equation is merely a convenient way of stating the relation between the induction in the iron core and the magnetising ampere-turns, and that it can in no respect be compared with Ohm's law for the electric circuit, which is founded upon the quality of constancy of the electrical conductivity of substances irrespective of the value of the current, whereas the multiplier μ is a variable quantity which can only be exactly determined by experiment.

Let us, now, in the place of the ring, consider a

straight iron bar, also uniformly wound with Si ampere-turns per centimetre of its length. The path of the magnetic lines is now partly through the iron and partly through the surrounding air. If the bar is short as compared with its section, the average induction in it will be determined almost entirely by the resistance of the air path. If, on the other hand, the bar is made longer and longer without limit, the resistance of the air path rapidly diminishes, and, if we consider an iron wire of which the length is about 300 times the diameter, the resistance of the air return path may be neglected, and the induction in the centre portions of the wire will be practically the same as in the case of the closed iron ring.

In order to get some idea of the distribution of a magnetic field in any particular case, it must be remembered (a) that all the lines of induction due to the exciting coils are closed lines, and (b) that every line will always choose the path of least resistance; or, in other words, that in any portion of a magnetic circuit the flux of induction will always be directly proportional to the magnetic difference of potential between the ends of the section, and inversely proportional to what has been called its resistance.

It is only in the case of the very simplest arrangements of magnetic circuits that the distribution of the induction can be correctly predetermined; but it should always be possible to get a general idea of the magnetic field likely to result from any particular arrangement of the magnetising coils.

If we place in the interior of a solenoid an iron core of comparatively small section, only a certain amount of the total flux of induction will pass through the iron; in fact, we have here an almost perfect electrical analogy in the case of a glass tube filled with mercury, through which there is a uniform flow of current. If we place in such a tube a length of copper wire, the

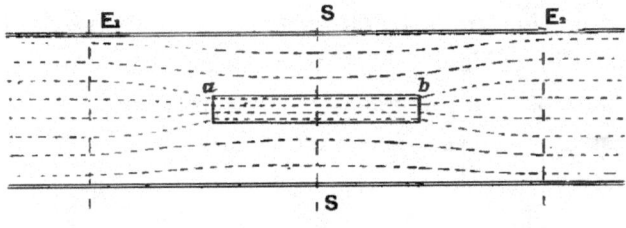

FIG. 1

lines of flow will crowd into the copper, on account of its greater conductivity, more or less in the manner indicated in fig. 1. If we assume a constant difference of potential between the two equipotential surfaces E_1 and E_2, we observe, after the introduction of the copper wire, that the current density in the mercury on the section S S is *less* than before, unless the copper rod $a\ b$ is very long and extends beyond the sections E_1 and E_2; and it is only in the case of such a long copper rod that the current density in the mercury will be uniform throughout the section S, and will bear the same relation

to the density in the copper as the conductivity of
the mercury bears to that of the copper.   The large
fall of potential in the neighbourhood of the ends of the
rod, where the lines of flow crowd together on entering
and leaving the copper, is the reason why the difference
of potential available for sending current through the
centre portions of the rod is relatively less in the case
of a short rod.

It is hardly necessary to return from this electrical
analogy to our starting point of the iron wire in a large
solenoid, the similarity between the two cases being
sufficiently evident.   This is not, however, the usual
way of explaining magnetic phenomena, but it is more
consistent with what has gone before, and possibly less
confusing than the theories of free magnetism and self
demagnetising effects of short bars, although, as a little
consideration will show, all the observed phenomena
may be equally well explained by assuming the rod
a b in our electrical analogy to produce a *back* E.M.F.
depending upon the number of amperes which pass into
and out of it at the ends.

7. **Magnetic Leakage. Effect of Satura-
tion.**—If, in the case of the uniformly wound iron ring
already referred to, an air gap had been introduced in
some part of the magnetic circuit, or even if the winding
on the ring had not been uniform, but had been confined to a small portion of the ring, a certain amount
of magnetism would have escaped laterally from the

surface of the iron, thus following an alternative path
through the surrounding air. The magnetism which
leaves the main path of the magnetic lines in this way,
and which rarely serves a useful purpose—but, on the
contrary, may often, as in the case of transformers,
be decidedly objectionable—is generally known as
leakage magnetism.

In most cases arising in practice the amount of
this leakage magnetism will be approximately proportional to the magnetising force, and indeed this is

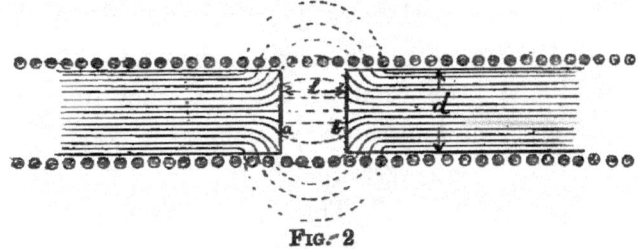

Fig. 2

probably true of any conceivable arrangement of the
magnetic circuit, provided the magnetisation of the iron
in the circuit is not carried above a value equal to about
half its saturation value; but as the iron becomes more
and more nearly saturated, the amount of the leakage
magnetism will, in certain cases, diminish, as will be
readily understood by reference to fig. 2.

Here we have the case of a uniformly wound helix
or solenoid with a discontinuous iron core. The ends
shown broken off in the figure must be considered either
as extending in both directions to a considerable

C

distance, or as being bent round so as to join, in which case the air gap $a\,b$ may be looked upon as a dividing slot in an otherwise complete iron ring.

When a current of electricity is sent through the coil, it is evident that, so long as the permeability of the air gap is less than that of the iron, a certain amount of the total magnetic flux will escape laterally from the iron and pass outside the coil in the manner indicated by the dotted lines. This 'leakage' magnetism will depend upon the length of the air gap; in fact, it may be experimentally shown that the ratio of the length l to the diameter d determines, within certain limits, the relation between the amount of magnetism which finds its way outside the winding and the amount which passes from iron to iron, through the air gap, within the coil. This relation is also found to remain practically unaltered notwithstanding variations in the magnetising current, unless the latter is increased to such an extent that the iron begins to show signs of saturation. When this stage is reached the *leakage coefficient*—which may be defined as the ratio of the total flux in the iron to that portion of it which crosses the air gap within the coil, and which hitherto has remained practically constant—begins to decrease, and continues to do so until, if it were possible to reach the point of saturation of the iron, there is every reason to believe that the leakage coefficient would become unity, whatever might be the dimensions of the gap; in other words, the magnetic lines would all pass within the coil between the two opposing faces of the iron core.

The reason why this point has been dwelt upon at some length, is not so much because it may be of practical importance when dealing with alternating current phenomena—indeed, questions of saturation of iron are not likely to arise in this connection—but principally because it is usual to estimate the value and distribution of leakage magnetism without giving proper consideration to the question of saturation. For instance, it is only permissible to compare a system of magnetic conductors to a similar system of electric conductors immersed in water, if it is clearly understood that the magnetisation of the iron is not carried beyond about half the saturation value; otherwise the analogy will, in all probability, be worthless, if not entirely misleading.

Those acquainted with the theories of magnetism will have no difficulty in explaining the observed reduction of the leakage magnetism with the higher magnetising forces in the special arrangement which we have just considered : it is a case for the application of Kirchhoff's *law of saturation*, in accordance with which we may say that, if we increase the magnetising ampereturns, the direction of the magnetisation, I, will, *at every point of the iron*, agree more and more nearly with the direction of the field, H, due to the coil alone, as the iron approaches the condition of saturation.

c 2

ALTERNATING CURRENTS

8. **Graphical Representation.**—The variations, both in strength and direction, of an alternating current may be accurately represented by means of a curve such as the one drawn in fig. 3. The lapse of time is measured horizontally from left to right, while the current strengths are measured vertically; hence the position of any point P on the curve indicates that, after an interval of time represented by the distance o *a*, the instantaneous value of the current is given by the length of the line *a* P; and further, since this measurement is made *above* the horizontal datum line OX, we conclude that the current is flowing in a *positive* direction. If the measurement had been made *below* this datum line, it would have been an indication that the current was flowing in a *negative* direction. The distance o *b* is evidently equivalent to the time of one complete period or alternation, after which the current must be considered as rising and falling periodically in identically the same manner. In practice, the time of one complete period is usually somewhere between the fiftieth and the hundredth part of a second,

the tendency in England at present being towards the
employment of comparatively low frequencies, i.e. about
fifty complete periods per second.

If n stands for the frequency, or number of complete
periods ($\sim$) per second, then the number of *reversals*
per second is $2n$, and the time of one complete alternation is $\dfrac{1}{n}$ seconds, which is numerically equivalent to
the distance o b in fig. 3.

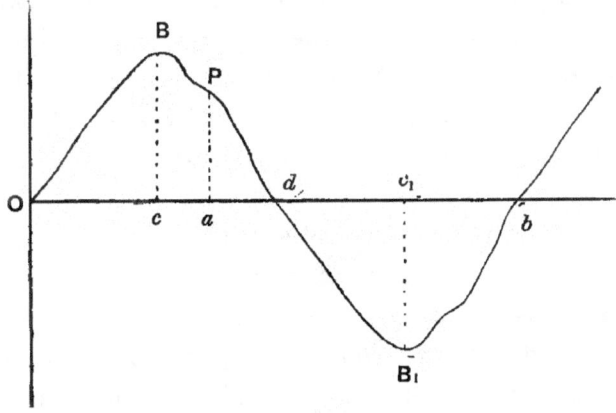

Let us suppose that the resistance of the circuit conveying the current represented in fig. 3 is r ohms;
then, if i is the current at any instant, the rate at which
work is being done in heating the conductors will,
at that particular instant, be expressed by the quantity
i^2r. Hence, if we wish to know the *average* rate at
which work is being done by an alternating current,

it will be necessary to calculate the mean value
of the *square* of the current and multiply this quantity
by the resistance r. In other words, if we imagine
a very large number of ordinates to be drawn, and
the average taken of the *squares* of all these ordinates, the product of this quantity by the ohmic resistance of the circuit will give us the watts lost in
heating the conductors. It follows, therefore, that when
we speak of an alternating current as being equal to a
certain number of amperes, we invariably allude to that
value of the current which, when squared and multiplied
by the resistance of the circuit through which it is
flowing, will give us the actual power in watts which is
being spent in overcoming the resistance of the conductors.

Thus it is the *square-root-of-the-mean-square* value
of an alternating current which is of primary importance,
and which, as far as *power* measurements are concerned,
enables us directly to compare a periodically varying
current with a continuous current of constant strength :
it is the product of this value of the current and the
corresponding value of the effective or resultant E.M.F.
to which it owes its existence which, in all cases, is a
measure of the power absorbed by the circuit.

The readings of nearly all commercial measuring
instruments for alternating currents depend upon the
$\sqrt{\text{mean square}}$ value of the current or E.M.F. ; and it is
only in exceptional cases that we require to know either
the *maximum* or the true *mean* value of an alternating

quantity, whether current or E.M.F., although it will be evident that questions of insulation must be discussed with reference to the *maximum* value of the E.M.F., which will be great or small according to the *shape* of the E.M.F. curve, even if the $\sqrt{\text{mean square}}$ value remains unaltered.

With regard to the *mean* value of a periodically varying quantity, it is hardly necessary to point out that this also depends upon the shape of the curve, and that it is by no means the same thing as the $\sqrt{\text{mean square}}$ value. It may, in fact, differ, even widely, from the latter ; but we are very little concerned with it at present. From an inspection of fig. 3 it will be seen that we have merely to take the average of the ordinates of the current wave, or divide the area of the curve opd by the length of the line od, in order to obtain the true mean value of the current.

It must, however, be clearly understood that it is the $\sqrt{\text{mean square}}$ value of an alternating current or E.M.F. with which we are principally concerned ; and when mention is made of amperes or volts in connection with variable currents, it is always this value which is alluded to, unless special reference is made either to the maximum, or the mean, or to an instantaneous value of the current or E.M.F.

9. **Clock Diagram.**—The best and simplest way of dealing with alternate current problems is, undoubtedly, by the aid of vector, or ' clock,' diagrams, such as the one shown in fig. 4. These diagrams were first intro-

duced by Thomson and Tait, and they are now exten-
sively used for the graphical solution of alternate
current problems.

Let the line OB, in fig. 4, be made equal in length
to the *maximum* value of the alternating current or
E.M.F. (cB or c_1B_1 in fig. 3) ; and let us suppose this

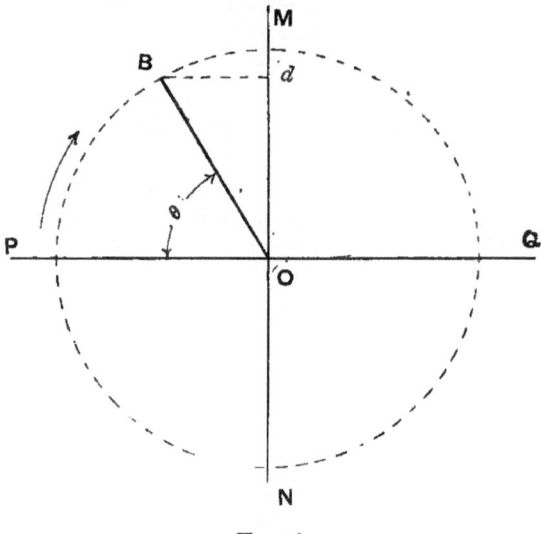

Fig. 4

line to revolve round the point O as a centre in the
direction indicated by the arrow. If, now, we consider
the projection of this revolving line upon any fixed
straight line—such as the vertical diameter MN of the
dotted circle—it will be seen that the speed of OB can
be so regulated that the length of this projection will,
at any moment, be a measure of the instantaneous

value of the variable current or E.M.F. It will also be evident that—since the current must pass twice through its maximum value, and twice through zero value, in the time of one complete period—the line OB must, in all cases, perform one complete revolution in $\frac{1}{n}$ seconds; where n is the frequency, or number of complete alternations per second. Also, in order that this diagram may give us all the information needed, it will be convenient to assume that all measurements, such as od, which are made *above* the centre o, correspond with the *positive* values of the variable quantity, whereas all measurements made *below* will apply to the *negative* values.

When the line OB (fig. 4) is vertical, its projection od is equal to it; we therefore conclude that the alternating quantity is at that moment passing through its maximum positive value. As OB continues to move round in a clockwise direction, od will diminish, until the point B has moved to Q, when od will be zero; after which it will again increase in length, but this time—since it is now *below* the line PQ -the flow of current, or the direction of the E.M.F., is reversed. At N the maximum *negative* value will be reached, only to fall again to zero at P; after which it rises once more to the *positive* maximum at M.

If the line OB revolves round o at a *uniform* rate, the point d will move to and from the centre o with a simple periodic, or simple harmonic motion. It follows that, if the length of the projection o*d* represents the variations of an alternating E.M.F., this E.M.F. must be understood to be rising and falling in a simple periodic manner; and since the length o*d* will now be proportional to the sine of the *time angle* θ, the shape of the wave (fig. 3) will be that of a curve of sines, the characteristic feature of which is that every ordinate, such as *a* P, will be proportional to the sine of its horizontal distance from o : this distance being now expressed, not in *time*, but in angular measure, it being, of course, understood that 360 degrees correspond to the time of one complete period.

It is on account of the almost general use of ' clock ' diagrams that intervals of time are frequently expressed in angular degrees; but, once the reason of this is understood, no confusion is likely to arise from the use of the expression.

10. **Addition of alternating E.M.F.s.**—In fig. 5 the vectors e_1 and e_2 must be thought of as representing the maximum values of two distinct alternating E.M.F.s of the same frequency of alternation, but with a *phase difference* proportional to the angle θ. These two E.M.F.s may be considered as being produced by a couple of alternators, having equal numbers of poles, and which are rigidly coupled together so as to be driven at the same constant speed. The armature windings of these machines being joined in series, the

question arises as to what will be the *resultant* E.M.F. of the combined machines

If the alternators are so coupled that the magnet poles of both of them are opposite the centres of the armature coils at exactly the same instant, the two E.M.F.s will be said to be *in phase*, and the lines oe_1 and

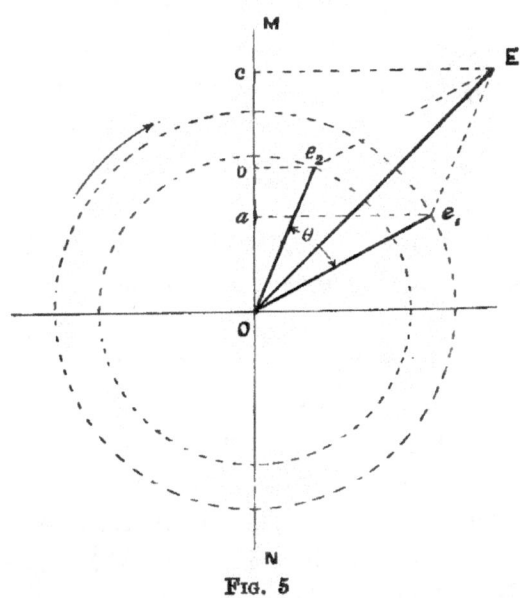

Fig. 5

oe_2 will coincide. As it is, we have supposed that the E.M.F. wave produced by one alternator reaches its maximum value at a time when the E.M.F. due to the other alternator has already passed its maximum, and is falling towards its zero value; the exact fraction of a period which corresponds to this *lag* of the one E.M.F.

behind the other being measured by the ratio which the angle θ bears to the complete circle.

The resultant E.M.F. generated will, at any instant, be equal to the sum of the projections oa and ob of the vectors e_1 and e_2 upon the line MN; and if we construct the parallelogram of forces in the usual way, and project the resultant OE upon MN, it will be seen that—since $bc = oa$—the projection oc of the vector E will, at any instant, be a measure of the total E.M.F. generated; it being understood that OE revolves round the centre O, at the same uniform rate as the two component vectors e_1 and e_2.

11. **Practical Application of Vector Diagrams.**—By means of such a diagram as the one just described, it is, of course, possible to add together any number of alternating E.M.F.s of the same periodicity; and in this way we are enabled to predetermine not only the *maximum* value of the resultant E.M.F., but also its instantaneous value at any particular moment. In order to do this, it is, however, necessary to assume that the component E.M.F.s are sine functions of the time; in other words, that the rise and fall of the E.M.F.s (including the resultant) is in accordance with the simple harmonic law of variation.

In actual practice this condition is not always fulfilled, and the E.M.F. waves produced by different alternators may be of various shapes; some having more or less pronounced peaks, corresponding to a comparatively large maximum value; others being flatter, and more rectangular in shape than the simple sine curve.

It is not often that we require to determine instantaneous values of an alternating current or E.M.F. It is generally sufficient—as already mentioned—if we know the $\sqrt{}$ mean square values, such as can be read off any alternate current ammeter or voltmeter.

Consider again two alternators, A and B, joined in series. They must still be supposed to have the same number of poles, and to be driven at the same speed; but the E.M.F. waves produced by the two machines, instead of being sine curves, may now be of any other shape, and the form of wave due to A may differ entirely from the wave due to B.

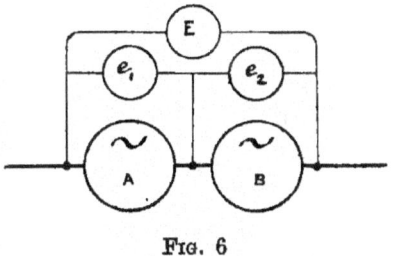

FIG. 6

Let us suppose three voltmeters to be connected as shown in fig. 6. These voltmeters must be such as may be used indifferently on alternating or direct current circuits; that is to say, they must measure the $\sqrt{}$ mean square values of the alternating volts. They may, for instance, be either electrostatic, or hot wire instruments; but they should not depend upon the electro-magnetic actions of coils having iron cores,

because the induction in the iron cores will depend
upon the *mean* value of the applied E.M.F., and the readings of such instruments will, therefore, be more or less
dependent upon the wave form of the impressed volts.

The voltmeters e_1 and e_2 will give us the volts due,
respectively, to the alternators A and B; whereas E
will measure the resultant volts at terminals.   If the
volts measured by E are equal to the arithmetic sum of
the volts e_1 and e_2, the two machines would be said to

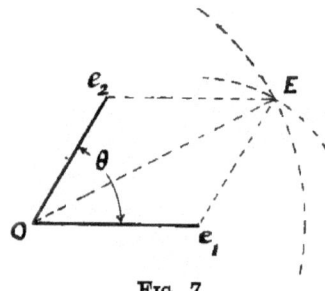

FIG. 7

be *in phase*, though it is
quite possible that the
maximum values of the
component E.M.F.s might
not agree, on account of
the waves produced by
the two alternators being
of different shapes.   As
a rule, the reading on E
will be *less* than the arithmetic sum of e_1 and e_2.
We will suppose these three values to be known.
From the centre o (fig. 7) describe a circle of radius
OE, the length of which is a measure of the volts
E.   Now draw oe_1 in any direction, to represent the
volts e_1.   From e_1 as a centre describe an arc of radius
e_1E, the length of which is proportional to the volts e_2;
it will cut the arc already drawn at the point E.   Join
OE, and complete the parallelogram.   Then the *angle
of lag* θ, between the vectors e_1 and e_2, is a measure of

what we must now understand as the phase difference between two alternating quantities, which are of the same frequency, but which do not necessarily follow the sine law of variation.

12. **Current Flow in Circuit without Self-Induction.**—Let us consider an electric circuit which is practically without self-induction, or electrostatic capacity. It may consist of a wire doubled back upon itself (in the manner adopted in winding resistance coils for testing purposes), or of glow lamps, or of a water resistance.

If an alternating E.M.F. is applied to the terminals of such a circuit, the current at any instant will be equal to the quotient of the instantaneous value of the E.M.F., divided by the total resistance of the circuit; or,

$$i_i = \frac{e_i}{r}$$

from which we see that the current wave will be of the same shape as the E.M.F. wave, and *in phase* with it ; a state of things which is the evident result of the fulfilment of Ohm's law : for there is no reason for supposing that Ohm's law is not equally applicable to variable as to steady currents ; it is only necessary to bear in mind that, in the case of variable currents, the *applied* E.M.F. and the effective or *resultant* E.M.F. in the circuit (to which the current is due) are not necessarily one and the same thing.   In the case under considera-

tion, of a circuit supposed to be without self-induction
or capacity, there is only one E.M.F. tending to produce
a flow of current, i.e. the E.M.F. supplied at the terminals
of the generator : the current will therefore rise and
fall in exact synchronism with the applied E.M.F.

The *power* at any instant will be equal to the product of the E.M.F. and corresponding current, or

$$W_i = e_i \times i_i.$$

By drawing the curves of current and E.M.F., and
plotting the product W_i for a number of ordinates, we
can readily obtain the *power curve*, the mean ordinate of
which will give us the average rate at which work is
being done by the current in the circuit.

The power at any instant being equal to $e_i \times i_i$, this
may also be written $\dfrac{e_i^2}{r}$ and $i_i^2 r$. Hence, if e and i stand
for the $\sqrt{\text{mean square}}$ values of E.M.F. and current, it
follows that $w = \dfrac{e^2}{r} = i^2 r = ei$, where $w =$ the mean ordinate of the power curve, which gives us the average
rate at which energy is being supplied to the circuit.
We therefore see that, in the case of a circuit which
may be considered non-inductive and without appreciable capacity, we have merely to take simultaneous
readings of the volts and amperes and multiply these
together in order to obtain the true watts going into
the circuit.

SELF-INDUCTION

13. **Coefficient of Self-Induction.**—The self-induction of a circuit for any given value of the current i passing through it is the total magnetic flux of induction through the circuit, *which is due to the current i.* If this current is a variable one, the self-induction will vary also. If the induction is produced by a coil of wire with an air core, it will vary in the same manner as the current; but if there is iron in the magnetic circuit, the law of variation of the self-induction is a less simple one.

The *coefficient of self-induction* of a circuit—generally denoted by L—is a quantity which it is occasionally useful to know; it is made much use of in the analytical treatment of alternate current problems, and leads to some simplifications when it is permissible to assume a complete absence of iron in the magnetic circuit; in which case L is constant, and may be defined as the amount of *self-enclosing* of magnetic lines by the circuit when the current has unit value. For the *coefficient* of self-induction takes into account the number of times that the total induction N is threaded through the circuit: for instance, if a circuit takes one or two turns

D

upon itself, it will be the same thing—as regards the
magnitude of the induced E.M.F.—as if it formed only
one loop, but enclosed two or three times the amount
of magnetism which we have called the self-induction.
Thus, if we consider a coil of wire of S turns, the self-
induction (if there is no iron) will be proportional to
the ampere-turns Si; but the coefficient of self-induction, L, will be proportional to S^2. If the coil of wire
had consisted of only a single turn, the coefficient of
self-induction might have been defined as the number
of magnetic lines which would be threaded through the
circuit if one (absolute) unit of current were flowing.
Let us denote this by N_o; then, if, instead of having
only one turn, the coil be supposed to consist of S turns,
the self-induction will be S times N_o, and the amount
of *self-enclosing* of magnetic lines will be N_oS^2. This
may be written NS, where N stands for the actual
number of lines threaded through a coil of S turns
when unit current is passing through the coil.

When there is iron in the path of the magnetic
lines, the coefficient of self-induction will no longer be
constant, but will depend to a certain extent upon the
value of the current. We can, however, still consider
L as a coefficient which, when multiplied by the
maximum value of the current, will give us the total
amount of *self-enclosing* of magnetic lines, if we put

$$L = \frac{NS}{I} \quad . \quad . \quad . \quad . \quad (6)$$

where I stands for the *maximum* value of the alternating current actually flowing in the circuit; and this definition of L applies equally well to the case of a circuit in which there is no iron, only, as N is then proportional to I, L will have a constant value which does not depend upon the current.

14. **Electro-motive Force of Self-Induction.**—Under the heading MAGNETIC PRINCIPLES, the relation between the magnetic flux and the induced E.M.F. has already been discussed. It was there shown that the E.M.F. generated in a coil is proportional to the *rate* of ' cutting,' or of enclosing or withdrawing magnetic lines. This relation between the changes of magnetism and the induced E.M.F. is, of course, in no wise altered if the magnetism producing this E.M.F. is due to the current flowing in the coil itself.

Let I be the maximum value of an alternating current passing through the coil, and N the total amount of magnetic flux produced by this current; then, since in one complete period the magnetic lines denoted by N are twice created and twice withdrawn, it follows that the *mean* value of the induced E.M.F., or E.M.F. of self-induction, will be

$$E_m = 4\text{N}Sn \quad . \qquad . \qquad . \qquad (7)$$

where n stands for the frequency of alternation, in complete periods per second; and this equation is true,

whatever may be the shape of the current wave.
Putting in the place of N its value LI/S, which may
be deduced from (6), and assuming the current wave
to be a sine curve, in which case the *mean* value of
the current, I_m, is equal to $\dfrac{2}{\pi}$ times its *maximum* value,
I, we obtain the expression

$$E_m = 2\pi n L \times I_m \qquad . \qquad . \qquad (8)$$

which is well known in connection with the analytical
treatment of the subject.

Since, in the above equation, L is measured in
absolute C.G.S. units, both E_m and I_m will, of course, be
expressed in the same units. If the E.M.F. is to be given
in volts and the current in amperes, we should have to
write

$$e_m = 2\pi n L i_m \times 10^9 \qquad . \qquad . \qquad (9)$$

but it is customary to omit the multiplier 10^9 and
thus make the *practical* unit coefficient of self-induction 10^9 times greater than the absolute C.G.S. unit.
To this practical unit coefficient of self-induction it
has been considered necessary to give a name, and it
is known as the *secohm* or *quadrant*, or, more recently,
the *henry*.

If, in order to furthur simplify the expression for
the electro-motive force of self-induction, we make the
assumption that there is no iron in the magnetic
circuit, in which case the induced E.M.F. will vary in

the same (simple harmonic) manner as the current, we may put

$$e = 2\pi n L i \quad . \quad . \quad . \quad (10)$$

where e and i are the $\sqrt{\text{mean square}}$ values of the E.M.F. and current; or

$$e = p L i \quad . \quad . \quad . \quad (11)$$

where p is a multiplier depending upon the frequency and the wave forms of the alternating current and E.M.F., and which, in this case, is equal to $2\pi n$.

15. **Effect of Wave Form on the Multiplier** p.—In order to understand of what practical utility the equation (11) may be, it will be necessary to consider the manner in which the multiplier p is affected by variations in the wave forms of the current and the resulting induced E.M.F. If, in the place of E_m in the fundamental equation (7), we put e_m for the mean induced E.M.F. in *volts*, this equation may be written:

$$e_m = \frac{4 N S n}{10^8}. \quad . \quad . \quad (12)$$

But, as it is not the *mean* value of the induced volts which we generally require to know, let us denote the ratio $\dfrac{\sqrt{\text{mean square}} \text{ volts,}}{\text{mean volts}}$ or $\dfrac{e}{e_m}$ by m, which we will call, for shortness, the *wave constant**; for, although it gives us no information regarding the actual shape of

---

* The reciprocal of Dr. Fleming's ' form factor ' (see his *Cantor Lectures*, 1896).

the E.M.F. curve, its value is all that we require to know in order that readings taken on a voltmeter may enable us to calculate the amount of the total magnetic flux through the circuit.   We may therefore write :

$$e = \frac{4mn\mathsf{N}S}{10^8}. \qquad . \qquad . \qquad (13)$$

But, by definition [see equation (6)], $\mathsf{N}S = L \times 10^9 \times \dfrac{I}{10}$, where L stands for the *practical* unit coefficient of self-induction, and I for the maximum value of the current in *amperes*; and, if we denote by r the ratio of the *maximum* current to the $\sqrt{\text{mean square}}$ current, then $I = ir$, and the equation (13) becomes :

$$e = (4mrn)Li, \qquad . \qquad . \qquad (14)$$

the quantity in brackets being the multiplier p of equation (11); and it is this quantity $(4mrn)$ which, in the future, must be understood to represent the numerical value of the symbol p.

   If both the current and E.M.F. waves are sine curves, of which the *maximum* ordinates may be considered equal to unity, the *mean* of all the ordinates will be equal to $\dfrac{2}{\pi}$, and the *square root of the mean of the squares* of the ordinates will be $\dfrac{1}{\sqrt{2}}$; hence $m = \dfrac{\pi}{2\sqrt{2}}$ and $r = \sqrt{2}$, which makes $p = 2\pi n$, as already stated.

16. Inductance.—In the expression $e = pLi$, the quantity pL is sometimes called the *inductance* * of the circuit; and it follows that, if we multiply the *inductance* by the current flowing, we obtain the E.M.F. of self-induction, e, or that component of the applied E.M.F. (exactly equal and opposite to e) which is required to overcome the inductance of the circuit and thus allow the current i to flow through it.

This way of looking at the question of the self-induction of a circuit conveying an alternating current has found favour on account of the resemblance between the expressions $\dfrac{e}{pL} = i$ and Ohm's law, $\dfrac{e}{R} = i$; but it must be remembered that—since, as already pointed out, the coefficient of self-induction, L, will sometimes depend upon the strength of the current—the inductance, pL, of a given circuit will not necessarily be constant, even with constant frequency. Also, in the case of Ohm's law, the component e of the impressed E.M.F. which produces the flow of current i through the *non-inductive* resistance R is exactly in phase with the current, and the product $e \times i$ *always* represents expenditure of energy; whereas the E.M.F. of self-induction being exactly one quarter period behind the current producing it, the power represented by the product of e and i in the expression $e = (pL)i$ is always equal to zero. This will be made clearer by what follows.

* This quantity (pL) is also called the *reactance*; the name *inductance* being then given to the coefficient of self-induction (L).

17. **Current Flow in Circuit having Appreciable Self-Induction.**—In order to get a better understanding of the whole question of self-induction in connection with alternating currents, let us consider an alternating current flowing in a circuit which has both ohmic resistance and inductance, as, for instance, a coil of wire of many turns, which, for the

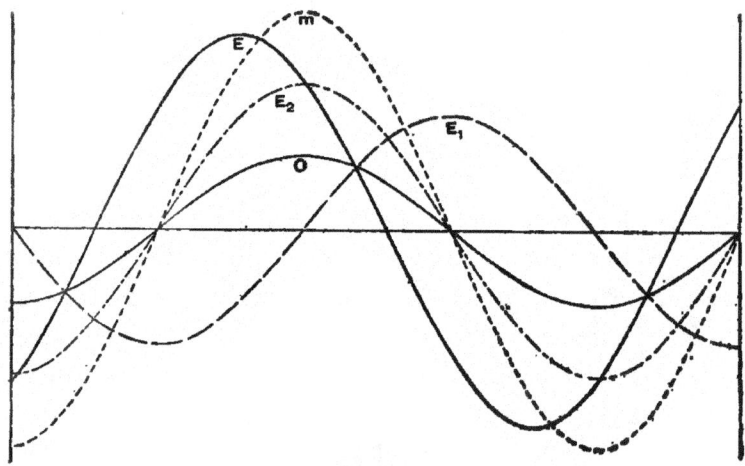

Fig. 8

present, we will assume, has no iron core. Such a current is shown graphically by the curve c in fig. 8, where intervals of time are measured, as usual, horizontally from left to right. The magnetism due to the current c will vary in amount and direction in accordance with the variations of the current. It may be calculated in the usual way for any given value of c, provided we know the length and cross-section or the *magnetic resistance* of the various parts of the magnetic circuit, and the number of turns of wire in the coil. Let the curve *m* represent the rise and fall of this magnetism.

Since the *induced* or *back* E.M.F. due to these variations in the magnetic induction will be proportional to the *rate of change* in the total number of magnetic lines threaded through the circuit, we shall have no difficulty in drawing the curve E_1, which represents the E.M.F. of self-induction. It is only necessary to remember that one hundred million C.G.S. lines enclosed or withdrawn per second will generate a mean E.M.F. of one volt per turn of wire in the coil.

Regarding the *direction* of the resulting induced E.M.F., a very simple rule will enable us to determine this with absolute certainty in every possible case. We have merely to bear in mind that the direction of this E.M.F. is *always* such as will tend to produce a flow of current *opposing* the changes in the magnetic induction. Thus, during the time that the magnetism is rising from its zero value to its maximum *positive* value, the induced E.M.F. will be *negative*; and all the while that the magnetism is falling from its maximum positive value to its maximum negative value, the induced E.M.F. will be in a *positive* direction; thus tending to produce a current which would prevent the fall, or check the rate of decrease, of the magnetic flux. It follows that the induced E.M.F. must always pass through zero value at the time when the magnetism threaded through the circuit is at its maximum. A graphical method of obtaining the curve of induced E.M.F. from the curve of magnetisation will be considered in due course.

With regard to the relation between the direction of the magnetising current c and that of the magnetic flux m, this is too well known to necessitate its being dwelt upon here. The analogy between the *forward* motion of a corkscrew and the *positive* direction (i.e. from S to N) of the magnetic flux, on the one hand, and the *right-handed* rotation of the corkscrew and the *clockwise* circulation of the current which produces the forward flow of magnetism, on the other hand, is a very useful one, and is of more general application than many others.

Having drawn the curve E_1 (fig. 8), which, it will be seen, lags, as already stated, exactly one quarter period behind the current wave, we are now in a position to determine the potential difference which must exist at the terminals of the circuit in question in order that the current c will flow through it.

Draw the curve E_2 to represent the E.M.F. required to overcome the ohmic resistance. It will be in phase with the current, because its value at any point is simply $C \times R$, where R stands for the resistance of the circuit. Now add the ordinates of E_2 to those of an imaginary curve exactly similar but opposite to E_1, and the resulting curve E will evidently be that of the impressed potential difference which, if maintained at the ends of the circuit under consideration, will cause the current C to flow in it. Thus we see how the relation between the impressed E.M.F. and the resulting current may be graphically worked out for any given case.

From a study of the curves in fig. 8 it is evident that the effect of self-induction is to make the current lag behind the impressed E.M.F. If the E.M.F. required to force the current against the ohmic resistance is small in comparison with the induced E.M.F., the lag will be very considerable; it cannot, however, exceed one quarter of a complete period, which limit is only reached when the E.M.F. of self-induction is so large, and the ohmic resistance of the circuit so small, as to render the E.M.F. required to overcome this resistance of no account.

In order to briefly sum up the principles governing the flow of an alternating current in a circuit having self-induction, we may say that the varying current produces changes of magnetism, which again produce a varying E.M.F., called the E.M.F. of self-induction. This, together with the E.M.F. already existing (and without which no current would flow), produces the *effective* or *resultant* E.M.F. By dividing the value of this resultant E.M.F. at any instant by the total ohmic resistance of the circuit, the corresponding current intensity is obtained. This condition *must* always be fulfilled, otherwise Ohm's law will not be satisfied.

18. **Graphical Method of Deriving the Curve of Induced E.M.F. from the Curve of Magnetisation.**—Let the curve *m* in fig. 9 represent (as in fig. 8) the variations in the magnetic flux through the circuit which we have been considering.

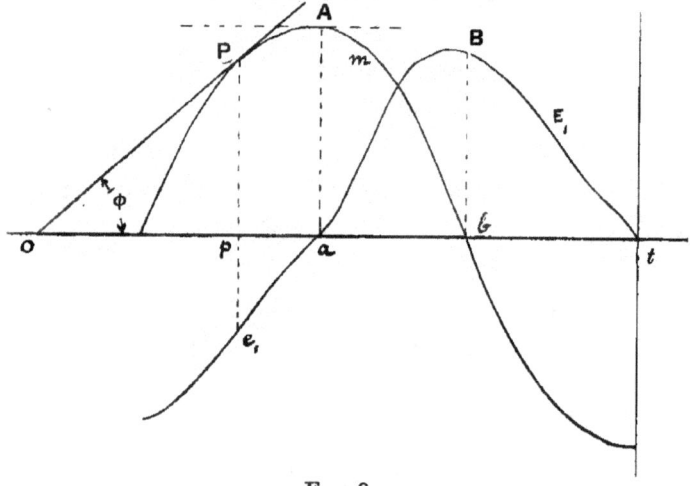

Fig. 9

Since the vertical distances above or below the datum line are a measure of the total number of magnetic lines passing at any moment through the circuit (the lapse of time being measured horizontally), it follows that the ' slope ' or ' steepness ' of the curve *m* will give us, for any point on the curve—and, therefore, at any particular instant—the *rate* at which the magnetic flux is changing. Thus, by drawing at any point P the tangent OP to the curve m, and then dividing the amount of the magnetic flux Pp by the lapse of time Op, we obtain a number which is proportional to the induced volts pe_1, and which enables us to plot the point e_1 of the curve E_1, which will be that of the induced E.M.F.

For instance, if the ordinate Pp represents 100,000 C.G.S. lines, and the distance Op the two-hundredth part of a second, then, if the circuit makes one hundred turns upon itself, the instantaneous value (pe_1) of the induced volts will be $10^5 \times 200 \times 100 \times 10^{-8} = 20$ volts, which must be plotted *below* the datum line, because this E.M.F. will be such as will tend to produce a current in a *negative* direction, i.e. such a current as would *oppose* the variation of the magnetic flux, which is *increasing* in amount.

At the point A, when the magnetism has reached its maximum positive value, the tangent to the curve is horizontal : the *rate of change* in the magnetism is therefore zero, and the point a will be on the datum line. The maximum value of E_1 will correspond to that point of the (falling) magnetisation curve which is steepest; but this is not necessarily the point b, where the curve m crosses the datum line. It is, however, interesting to note that the line bB always divides the curve aBt into two equal areas, abB and bBt, whatever may be the shape of the magnetisation curve. If the latter is a curve of sines, the curve E_1 will also be a curve of sines, of which the maximum ordinates will correspond with the zero values of the curve m.

To those unacquainted with the differential calculus it may not be quite clear why the 'slope' of the tangent OP is a measure of the rate of growth or of decrease

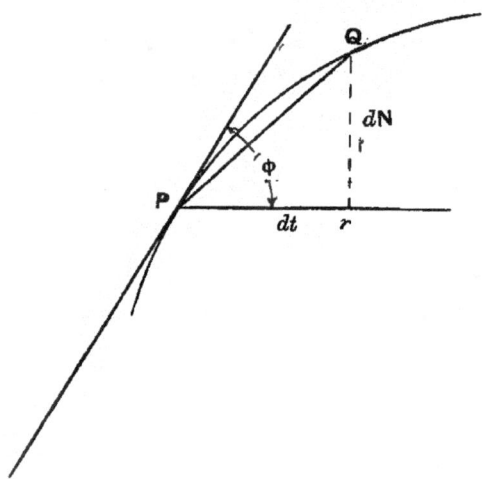

FIG. 10

of the number of magnetic lines threaded through the circuit. Let us therefore consider (see fig. 10) a magnified portion of the curve m of fig. 9, at the point P.

Let Q be another point on the curve m, situated at a small distance from P. Join PQ: then, since the amount of the magnetic flux through the coil has increased by an amount $d\mathsf{N}$ (represented by the distance Qr) in the time dt, it follows that the average *rate* of increase of the magnetisation, while changing from its value at P to its value at Q, will be given by the fraction $\frac{\mathrm{Q}r}{\mathrm{P}r}$ or $\frac{d\mathsf{N}}{dt}$, which will therefore be a measure of the mean value of the induced volts during the time taken by the magnetism in growing from P to Q. But the fraction $\frac{\mathrm{Q}r}{\mathrm{P}r}$ is the trigonometrical tangent of the angle QPr which the line PQ makes with the horizontal line Pr. If, therefore, we imagine the point Q to move nearer and nearer to the point P, the limiting position of the line PQ will be the tangent to the curve at the point P; and the ordinate pe_1 of the curve of induced E.M.F. (fig. 9) will therefore be proportional to the *trigonometrical tangent of the angle* (ϕ) *which the tangent to the curve at the point* P *makes with the horizontal datum line.* This being the differential coefficient of N with respect to t, it follows that in order to obtain the curve of induced E.M.F. we have to differentiate the curve of magnetisation. Thus the instantaneous value of E_1 will be given (in volts) by the expression

$$e_1 = \frac{S}{10^8} \frac{d\mathsf{N}}{dt}. \qquad . \qquad . \qquad . \quad (15)$$

where S stands, as before, for the number of turns in the coil, or the number of times which the magnetic flux N is threaded through the circuit. On the assump-

tion that there is no iron in the magnetic circuit, this
equation (15) can be written:

$$e_1 = L\frac{dc}{dt} \quad . \qquad . \qquad . \qquad . \quad (16)$$

where L is the constant coefficient of self-induction of
the circuit, as previously defined; and the symbol $\frac{dc}{dt}$
denotes the rate at which the current is changing in
strength at the particular moment in question.

19. **Mean Power of Periodic Current.**—
Returning to our study of the current flow in a circuit
having both ohmic resistance and self-induction—such
as the one to which the curves of fig. 8 apply—the
question arises as to what is a true measure of the
power being supplied to the circuit.

If we connect a voltmeter across the terminals of
the circuit, and multiply the reading on this voltmeter
by the actual current in amperes, we shall obtain a
number representing what are sometimes called the
apparent watts, but which will not be a measure of the
power actually being supplied to the circuit. The true
power at any moment will be given by the product
of the instantaneous values of current and impressed
potential difference; and the mean value of all such
products, taken during the time of one complete period,
will be the quantity which we require to know.

In fig. 11 the power, or watt curve has been drawn;

it is obtained by multiplying together the corresponding ordinates of the curves of impressed potential difference E, and of the current C.

Since the current *lags behind* the potential difference, it follows that during certain portions of the complete period the simultaneous values of E and C will be of

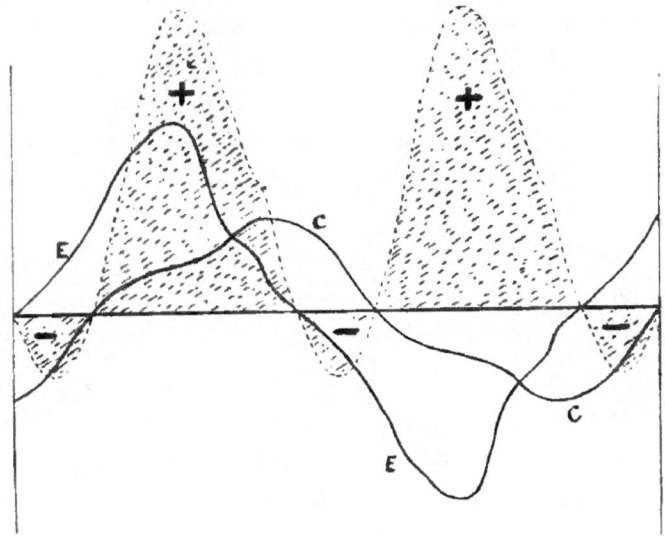

Fig. 11

opposite sign ; that is to say, the current will be flowing *against* the impressed E.M.F. : the work done will therefore be *negative*, and these ordinates of the watt curve will have to be plotted *below* the datum line. This negative work (which is equal to the area of the shaded curve below the datum line) may sometimes

E

almost equal the *positive* amount of work done, in which case the current is practically *wattless*—i.e. the amount of energy put into the circuit during one quarter period is given back again during the next quarter period.

The proper understanding of this state of things must of necessity present some difficulties to those unacquainted with alternating currents; but it must be remembered that whenever a circuit has appreciable self-induction, the current flowing in it behaves in every respect as if it had appreciable momentum. It will not grow simultaneously with the applied E.M.F., nor will it immediately fall to zero when the latter is removed.

We know that, in the case of a flywheel, in which the friction of the bearings and the air resistance may be considered negligible, work has nevertheless to be done in order to bring it up to its normal speed. The energy thus spent is stored up in the revolving wheel, and if we remove the force which we have applied in order to bring it up to speed, exactly the same amount of energy which we have put into it is now available for doing work, and it will all be given back again by the time the flywheel is brought to rest.

Almost exactly the same thing occurs in the case of a magnetic field. No work need be done in *maintaining* it, but energy was spent in *creating* it, and this energy will all be given back again to the exciting circuit by the withdrawal or removal of the magnetic field. It is for this reason that the product of the magnetic flux N by the number of turns S which the circuit makes upon itself is sometimes called the *electro-magnetic momentum* of the circuit.

Returning to the consideration of the curves of fig. 11, we see that the total amount of work done during one complete period is equal to the area of the two shaded curves marked $+$, *less* the area of two of the shaded curves marked $-$: and the mean power supplied to the terminals of the circuit will be given us by the average ordinate of this dotted *watt* curve, due attention being paid to the *sign* of the instantaneous power values.

It has been shown (p. 42) that the curve E may be considered as being built up of two components, one exactly in phase with the current C, which we have called the resultant or effective E.M.F. and which is equal to the product of O by R, the resistance of the circuit; the other exactly equal and opposite to the E.M.F. of self-induction, which will be of such a shape and in such a position that the mean product of its ordinates with those of the current wave C will be equal to zero. It therefore follows that if we multiply the ordinates of the current wave by those of the curve of effective E.M.F. (in phase with C) we shall obtain another watt curve of which the area will be exactly equal to that of the one drawn in fig. 11. The mean of all its ordinates will be equal to $c \times e_2$, where c and e_2 are the $\sqrt{\text{mean square}}$ values of the current and effective E.M.F. This product may, of course, also be written $c^2 R$.

20. **Vector Diagram for Current Flow in Inductive Circuit.**—Instead of drawing the actual wave forms of the current and various component E.M.F.s as in fig. 8, we may represent the relative positions and magnitudes of these quantities by means of a vector diagram which, notwithstanding its extreme simplicity, will enable us to compound the various E.M.F.s, and make all necessary power calculations.

It will be found that *whatever may be the shape of the current wave C in* fig. 8, the impressed volts e will always be equal to the square root of the sum of the squares of the induced volts e_1 and the effective volts e_2. If, therefore, we compound these E.M.F.s in the manner described in § 11 (see fig. 7), the phase difference between e_1 and e_2 will be 90°; and this represents the lag of a quarter period which, we have seen, exists between the induced volts and the current (which is in phase with e_2).

In fig. 12 oc is the current; oe_2 the effective component of the impressed potential difference (in phase with c); oe_1 the induced E.M.F. (drawn at right angles to oe_2), and oe the impressed potential difference, which is obtained by compounding the force oe_2 with the force oe'_1, exactly equal and opposite to e_1. The angle θ is the *phase difference* between the impressed E.M.F. and the resulting current.

Since e_2oe is a right-angled triangle, it follows that the *power* supplied to the circuit $(c \times e_2)$ can be written $c \times e \cos \theta$. But $c \times e$ is what we have already called the *apparent watts*. Hence it follows that the *real watts*= the *apparent watts* $\times \cos \theta$, and this is the definition of the *angle of lag* between current and impressed E.M.F. which is the most generally useful. It is this multiplier ($\cos \theta$) to which Dr. J. A. Fleming has given the name of *power factor*.

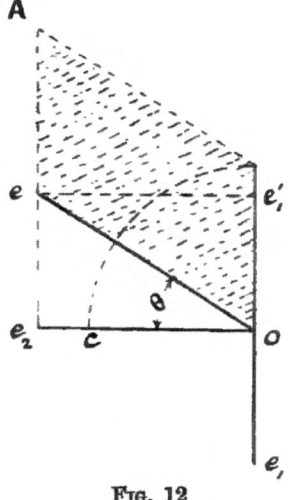

Fɪɢ. 12

In order to get a graphical representation of the power supplied to a circuit in which the current c lags behind the impressed volts e by an amount equal to the angle θ, we have simply to move round one of the vectors, let us say oc, through an angle of 90°, and then construct the parallelogram oA, the area of which will be a measure of the average value of the true watts; for it is evident that this area will always be equal to $oc \times oe_2$.

If the reader has carefully studied the curves of fig. 8, and understood the diagram just described, it

will hardly be necessary to point out that however much
we may increase the self-induction of the circuit—
and therefore the volts which must be supplied to the
terminals in order to keep the current constant—the
work done in a given time will remain the same,

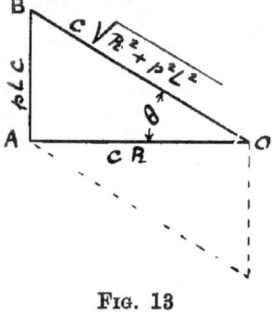

provided the resistance of the
circuit remains unaltered.

21. **Impedance.**—In the
triangle OAB (fig. 13), which
is a reproduction of the triangle oe_2e of fig. 12, the volts
OA may be written cR, and
$AB = pLc$ (see § 14, p. 37), hence
OB, which is the hypotenuse

Fig. 13

of a right-angled triangle, is equal to $c\sqrt{R^2 + p^2L^2}$. If,
now, we divide each of these quantities by c, we may
write :

OA ∝ R, the *resistance* of the circuit,

AB ∝ pL, the *inductance* of the circuit,

OB ∝ $\sqrt{R^2 \times p^2L^2}$, the *impedance* of the circuit,

and if we divide the amount of the impressed potential
difference by what has been called the *impedance*, we
shall obtain the value of the current flowing.   But the
expression $\dfrac{impressed\ volts}{amperes\ in\ circuit} = impedance$ is only a convenient way of stating the relation between E.M.F. and
current ; for it must not be forgotten that the impedance

is merely a multiplier which is not constant for a given circuit, but depends, among other things, upon the frequency and wave form of the impressed E.M.F.

22. Practical Measurement of Power in Inductive Circuit. Three Voltmeter Method. In the case of the inductive circuit to which fig. 12 applies, the induced volts e_1 cannot be directly measured ; but we can measure e, R, and c, hence we can calculate e_2 and draw the diagram fig. 12.

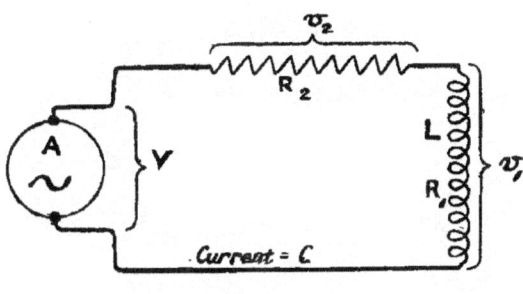

FIG. 14

The power in an inductive circuit can, however, be determined without knowing its resistance. Let L in fig. 14 be the inductive circuit, in which it is required to measure the power absorbed, and let it be connected up to the alternator A through the *non-inductive* resistance R_2. Measure the volts V, v_1 and v_2 across the alternator terminals, circuit terminals, and resistance R_2 respectively, and also the current c. Then, in fig. 15, draw oc to represent the current, and ov_2 for the

volts v_2 in phase with c.   From the point v_2 describe an
arc of radius equal to v_1, and from o another arc equal
to v.   Join their point of intersection v, and the point
o.   Then ov represents the potential difference at the
alternator terminals, and ov_1,—obtained by completing
the parallelogram in the usual way—represents the
potential difference at the terminals of the circuit L.

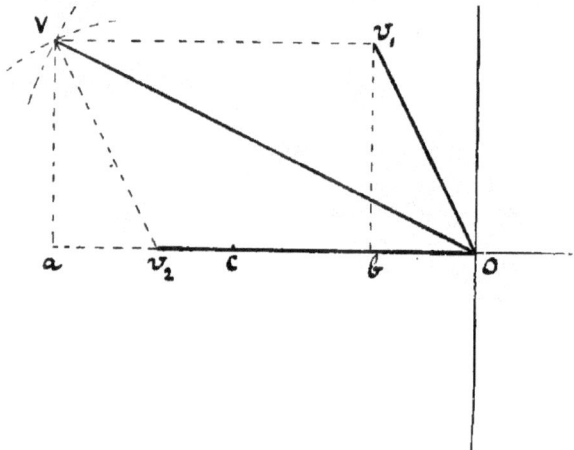

FIG. 15

The total power supplied at the alternator terminals
is equal to oc $\times$ the projection of ov on oc $=$ oc $\times$ oa.   The
power absorbed by the resistance R_2 is oc $\times$ ov_2; the
difference—which is the power absorbed by the resis-
tance R_1 of L—being equal to oc $(oa - ov_2)$, or to oc $\times$ the
projection of ov_1 on oc.   This *effective* voltage (ob) in the
circuit L may be written $\dfrac{ov^2 - ov_1{}^2 - ov_2{}^2}{2(ov_2)}$, as a study of

the diagram will show. Hence the power supplied to
the circuit L is equal to $c \times \dfrac{V^2 - v_1{}^2 - v_2{}^2}{2v_2}$, or if, instead of
measuring c, we prefer to measure the resistance R_2, this
expression becomes

$$w = \frac{V^2 - v_1{}^2 - v_2{}^2}{2R_2} \qquad . \qquad . \qquad .(17)$$

which is well known in connection with the *three volt-meter* method of measuring the power supplied to an
inductive circuit.

23. **Effect of Iron in Magnetic Circuit.
Eddy Currents.**—If, in our choking coil—which so
far has been supposed to have a core of non-conducting
material—we now introduce an iron, or, indeed, any
metal core, there will be currents generated in the mass
of the metal owing to the changes in the magnetic flux
passing through it. These eddy currents, being in phase
with the induced E.M.F., will tend to oppose the
changes in the magnetism.   Their mean demagnetising
tendency for a given shape and size of core will be
proportional to the induction, the frequency, and the
specific conductivity of the metal. By laminating or
dividing the metal core in a direction parallel to the flow
of magnetism, and insulating the adjacent plates or
wires with thin paper or varnish, these eddy currents
may be reduced to an almost negligible amount.

In transformers and other alternating current
apparatus, a very usual thickness for the sheet iron

stampings is ·014 in. Anything thicker than this, especially with the higher frequencies, would lead to quite appreciable losses ; but, on the other hand, there is little or no advantage to be gained by making the laminations less than ·012 inch in thickness.

The following empirical formula may be used for approximately determining the watts lost owing to eddy currents in the cores of transformers or choking coils, on the assumption that the amount of lateral leakage of magnetism is practically negligible :

$$\text{Watts per lb.} = \frac{1 \cdot 5}{10^{10}} t^2 n^2 B^2,, \qquad . \qquad . (18)$$

where $t =$ the thickness of plates in inches,
         $n =$ the frequency in periods per second,
         $B,, =$ the maximum induction in C.G.S. lines
                 per square inch.

That this formula is not theoretically correct will be evident from the fact that the loss of power is assumed to be proportional to the square of the maximum induction B, which must depend, to a certain extent, upon the *shape* of the applied potential difference wave. Strictly speaking, the eddy current losses should be expressed in terms of the E.M.F. of self-induction generated in the coil itself, for they will be proportional to the square of this quantity.

Although the power wasted by eddy currents is actually spent in heating the iron which lies in the path

of the magnetic lines, it will, of course, have to be put
into the magnetising circuit in the form of electrical
energy ; that is to say, there will be a certain component of the total current required solely to balance
the eddy currents, and which will therefore be unavail-
able for magnetising the core.   Thus, in the case of a
choking coil through which the current is kept constant
whatever may be the eddy current loss, the E.M.F. of
self-induction will be less when eddy currents are
present than it would be if these could be entirely
eliminated.   If, on the other hand, the volts across the
terminals of the choking coil be kept constant, the
current in the coil will increase in such a manner as to
neutralise the demagnetising effect of the eddy currents,
and thus leave the induction (and consequently the
induced E.M.F.) practically the same as it would be
if there were no eddy currents in the iron core.   In
fact, the effect of the eddy currents upon the magnetising coil will be almost exactly the same as if
current were taken out of a secondary coil, as in the
case of transformers (which will be discussed in due
course), and there will be a current component added
which will be in phase with that component of the
applied potential difference which balances the induced E.M.F.   It follows that if we multiply these
two quantities together we shall obtain the measure
(in watts) of the power wasted by eddy currents in
the iron core.

24. **Hysteresis.**—Although no energy is lost in producing changes of magnetism when the path of the magnetic lines is through air only, as soon as iron is introduced to convey and increase the amount of this magnetism, power is spent in magnetising and demagnetising the iron core, owing to what Professor Ewing has called *Hysteresis*.

This second source of loss, it should be clearly understood, is quite independent of the eddy current loss, and in no wise depends upon whether the iron is laminated or not.

It is well known that all iron, even the softest and purest, retains some magnetism after the magnetising force has been removed. By applying a magnetising force in the opposite direction this *residual* magnetism is destroyed, and the magnitude of this force—or, in other words, the amount of work which has to be done to withdraw this magnetism—depends upon the quality of the iron. Soft annealed wrought iron retains most magnetism; but, on the other hand, it parts with it more easily than the harder qualities of iron and steel, and for this reason requires the least expenditure of energy to carry it through a given cycle of magnetisation.

In fig. 16 is clearly shown the effect of hysteresis, which is to make the changes in the magnetism *lag behind* the changes in the exciting current. This curve is the ordinary magnetisation curve of a sample of transformer iron, and it gives the exciting force and corresponding induction for a complete cycle. The magnetising current is measured horizontally, on each side of the centre line, and may be considered *positive* when to the right, and *negative* when to the left of the latter. The length of the ordinates above or below the horizontal centre line is a measure of the induction in the iron for any particular value of the exciting current; the direction of the magnetism will be *positive* if the ordinates are measured *above* this datum line, and *negative* if measured *below*.

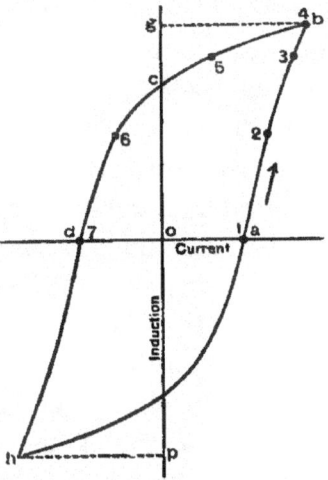

Fig. 16

Starting at *a*, and following the curve in the direction of the arrow, we see that it has required a certain definite *positive* current indicated by the distance *oa* to entirely withdraw all the *negative* magnetism. From *a* to *b* both current and induction rise to their maximum values; but from *b* to *c*, although the current falls from its greatest value to zero, the induction changes very little; in fact, the residual magnetism at *c* is not much below the maximum amount. However, as soon as a negative exciting force is applied, the magnetism decreases rapidly until when d is reached, and all the *positive* magnetism has been removed, it will be seen that the current has reached a *negative* value od, exactly equal to its *positive* value at a.

If the sample experimented upon were a closed ring which had been magnetised to what may be considered the saturation point, the *residual* magnetisation (represented by the distance oc) might be expressed as a percentage of the saturation value, and it would then depend only upon the magnetic properties of the iron used in the test.

Again, on the assumption that the point b represents the practical limit of magnetisation, the distance oa—which is a measure of the magnetising force required to withdraw the whole of the residual magnetisation—will also have a definite value depending upon the quality of the iron under test, and it will not even depend upon the shape of the sample, but only upon the limiting value of the magnetisation. For these reasons we are justified in giving to the demagnetising force, defined as above, a name of its own; and Dr. Hopkinson has called it the *coercive force*. Instead of being expressed in terms of the current passing through the exciting coil, the *coercive force* may also be defined as the demagnetising field, H (see § 5, under MAGNETIC PRINCIPLES) required to neutralise the residual magnetism, and this is the usual definition of the term.

In fig. 17 the curve of magnetisation, m, has been drawn in the same manner as in fig. 8 (p. 40), its maximum ordinate being scaled off fig. 16. The current curve c has then been plotted from measurements taken on the curve, fig. 16, which gives the current intensity corresponding to any particular value of the induction.

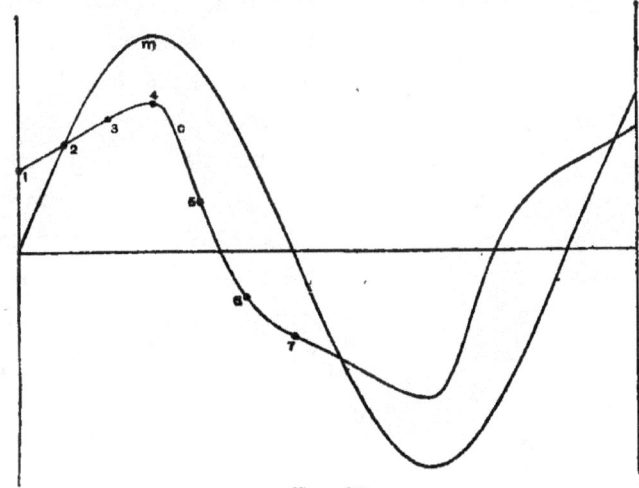

FIG. 17

In order to show quite clearly the manner in which the current curve c is derived from fig. 16, a few corresponding points on the two curves have been marked and numbered. Thus, in fig. 16, the distance of the point 5 from the vertical centre line gives us the magnetising current for a certain definite value of the induction. This magnetising current is represented in fig. 17 by the distance of the point 5 from the horizontal datum line, and the ordinate of the magnetisation curve which passes through this point will be found to be exactly equal to the ordinate of the corresponding point in fig. 16.

By comparing fig. 17 with fig. 8, it will be seen that the introduction of an iron core in the place of an air core (even if eddy currents are assumed to be absent) has considerably distorted the current wave, which now no longer rises and falls in synchronism with the magnetisation wave, but, in general, *precedes* it ; although it should be noted that — except when eddy currents are present — the *maximum* values of the induction and current still occur at the same instant.

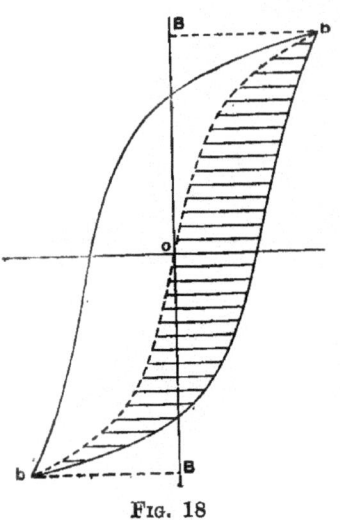

Fig. 18

25. 'Wattless' and 'Hysteresis' Components of the Magnetising Current.—The dotted centre line *hob*, in fig. 18, may be taken as representing the relation between exciting current and magnetism on the assumption that hysteresis is absent. Hence it follows that, at any point in the real cycle, the portion of the current which may be considered as doing

the work against hysteresis is indicated by the length
of the current line comprised between the dotted line
and the outside curve. This current will be *positive*
when measured to the right of this new centre line,
and *negative* when to the left. The short horizontal
lines in the figure represent the successive *positive*
values of this hysteresis current during the change of

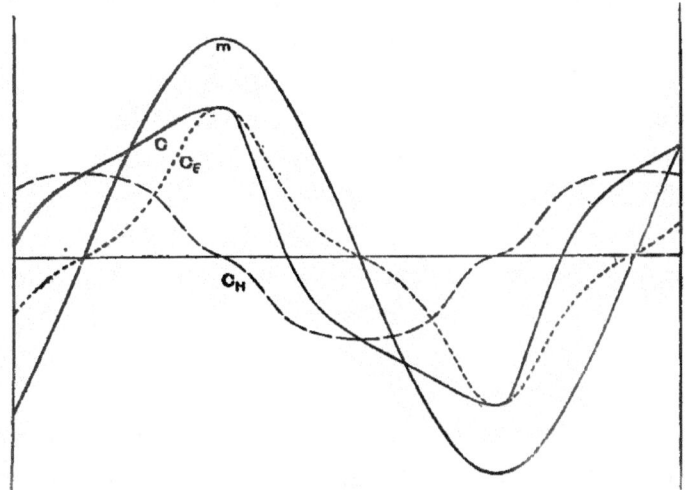

FIG. 19

the magnetism from its negative to its positive
maximum. In fig. 19 the curve of magnetisation is
marked m as before. c_E is the imaginary exciting
current for the assumed condition of no hysteresis in
the iron, it is plotted from measurements taken on
hob in fig. 18, while c_H is what we may call the
hysteresis component of the current, and is also

F

drawn from measurements made on fig. 18. The
resultant or true magnetising current is, at every
point, equal to the algebraic sum of these two.

26. **Power lost owing to Hysteresis.**—
The impressed potential difference required in order
that the current c may flow in the circuit will consist
—as already fully explained—of two components : one
equal to c × R, which is required to overcome the ohmic
resistance; and the other exactly equal and opposite
to the E.M.F. induced in the circuit owing to the rise
and fall of the magnetism, m. It is evidently the product of this last component of the impressed potential
difference and the current c which will give us the
power supplied to the circuit on account of the hysteresis
losses in the core : the latter, it should be added, being
due to a kind of internal or molecular friction which
has the result, as in the case of the eddy current losses,
of heating the iron core.

Now, the phase of the induced E.M.F. being
always, as previously explained, exactly one quarter of
a period behind that of the magnetism, it follows that
the current component c_E is wattless; but if we take the
ordinates of c_H and multiply them by the ordinates of
the induced E.M.F. curve (not shown in fig. 19), we
obtain the rate (in watts) at which work is being
absorbed through hysteresis. Note also that if the
frequency remains constant, the induction B (fig. 18)
is a measure of the induced E.M.F., and that the lost

watts are, therefore, directly proportional to the product of c_H and **B**, or to the area of the curve fig. 18.

There is another way of proving that the area of the hysteresis curve is a measure of the energy spent in carrying unit volume of the iron through one complete cycle of magnetisation.

Let us suppose that in fig. 20 we have plotted the magnetising force in ampere-turns (Si) per unit length of the iron core, and the corresponding induction **B** in C.G.S. lines per unit cross-section of the core, for one complete cycle of magnetisation. Consider the right-hand portion of the curve, from n to b.

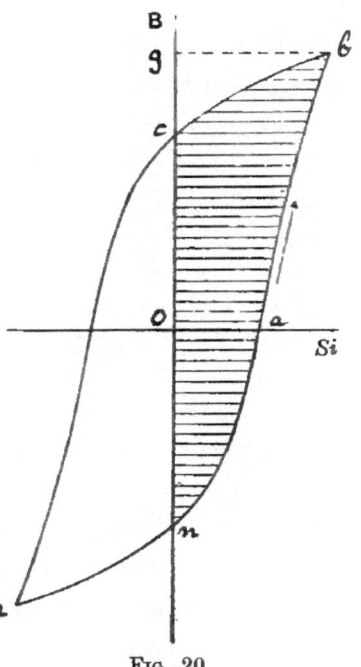

Fɪɢ. 20

The magnetising ampere-turns, Si, have risen from zero value at n to their maximum positive value gb; while the magnetic induction **B** has changed from its negative *residual* amount on, to its positive maximum amount og. Let us suppose this change to have taken place in the time t seconds. Then, if A is the cross-section of the core and l its length, the mean induced back E.M.F. due to the change in the magnetism will evidently be:

$$e \quad \text{(in volts)} = \frac{A \times (ng) \times Sl}{t \times 10^8}$$

and since the *quantity* of electricity which has been moved *against* this E.M.F. is equal to the mean value, i_m, of the current between its value at n and at b, multiplied by the time t, it follows that the *work* done is equal to:

$$\frac{Al}{10^8} \times (ng) \; Si_m$$

or to $\dfrac{Al}{10^8} \times$ the area *nabg*.

Thus we see that a distinctly appreciable amount of energy has been spent in raising the magnetic induction from its value $-on$ to its maximum value $+og$; if, now, we destroy this magnetism, the whole or a part of the energy required for its production will be restored to the circuit, owing to the fact that the E.M.F. induced by the withdrawal of the magnetism will be in the opposite direction to what it was before, i.e. it will now *help* the exciting current instead of opposing it.

If hysteresis were absent, the magnetism would fall in the same way as it rose, thus restoring the whole of the energy to the circuit. But in fig. 20 the magnetism falls from b to c along the curve bc, during the withdrawal of the whole of the magnetising force, thus only restoring to the circuit the amount of energy represented by the area *gbc*. It follows that the amount of energy lost during one half period—that is to say, while the current rises from zero to its maximum value, and falls again to zero—will be equal to $\dfrac{Al}{10^8} \times$ the shaded area *nabc*.

Exactly the same arguments apply to the left-hand portion of the curve (*chn*), which is a repetition of the curve *nbc*; and, since the product *Al* is equivalent to the volume of the iron core, we may write : work expended per cycle

$$= \frac{\text{volume of iron in core}}{10^8} \times \text{area of hysteresis curve; and}$$

since we have taken the back E.M.F. in volts and the current in amperes, the above amount of work will be expressed in *joules*, or practical units.

But, since $\mathsf{H} = \dfrac{4\pi}{10} \, Si$, instead of plotting B and Si, we may plot B and H, and, in order to express the energy lost in absolute C.G.S. units, we must convert joules into ergs (1 joule $= 10^7$ ergs) and take all measurements in centimetres; then ergs per cycle per cubic centimetre $=$

$$\frac{\text{area of hysteresis curve}}{4\pi} \quad . \quad . \quad . \quad . \quad . \quad (19)$$

The work absorbed by hysteresis depends very considerably upon the quality of the iron. In any one sample it is approximately proportional to the 1·6th power of the limiting induction, if the whole range of magnetisation is considered. For the straight part of the magnetisation curve which corresponds to the low inductions, as used in nearly all alternate current apparatus, it has been pointed out by Professor Ewing that the losses do not increase so rapidly, but more nearly as the 1·5th power of the induction; and it is for this reason that the law $B^{1·55}$ is adopted in the formula given below.

The number of alternations per second does not appreciably influence the hysteresis loss per cycle, excepting when the latter is very slowly performed; hence the watts lost per pound of iron for any particular maximum value of the induction will be proportional to the frequency. The following formula applies to a good quality of transformer iron and gives the watts lost per lb.

$$w = \frac{6·8}{10^{10}} n B_{,,}^{1·55} . \qquad . \qquad . \quad (20)$$

where $B_{,,}$ is the maximum value of the induction in C.G.S. lines per square inch, and n is the frequency, or number of complete periods per second.

If B stands for the induction in lines per square centimetre, the above formula becomes :

$$w = \frac{135}{10^{10}} n B^{1·55} \qquad . \qquad . \quad (21)$$

To obtain the watts lost per cubic inch, multiply by ·28.

The curve, fig. 21, gives the relation between the

limiting values of the induction and the watts lost per lb., at a frequency of 100, for the particular sample of transformer iron to which the above formulæ apply.

Curve giving hysteresis loss in watts per lb. of good quality transformer iron at a frequency of **100**

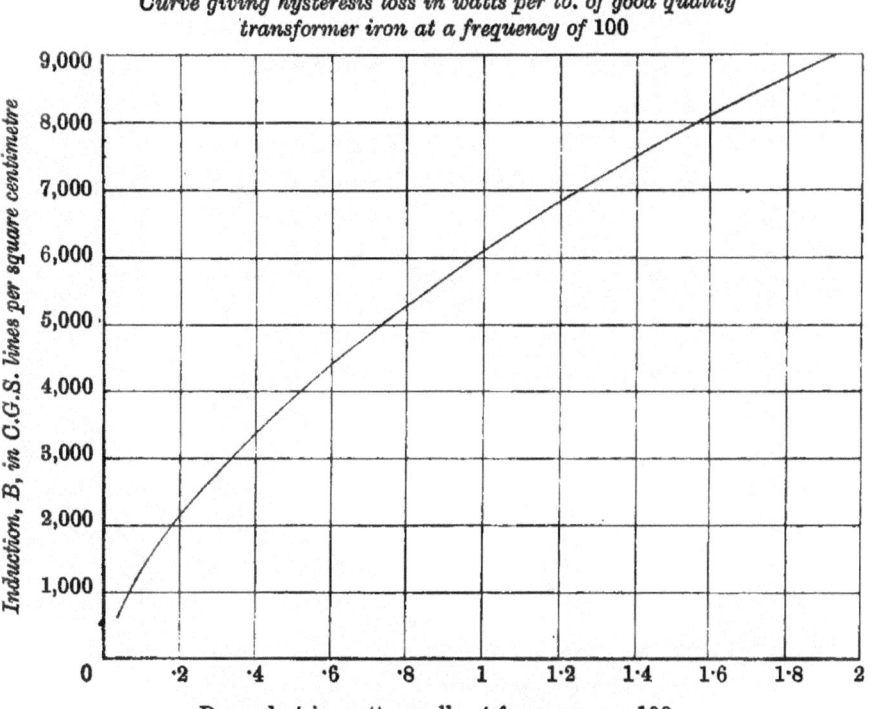

Power lost in watts per lb. at frequency $n = 100$

FIG. 21

An induction of 2,500 lines per square centimetre (approximately 16,000 lines per square inch) and a frequency of 100 are often taken as standard conditions for purposes of comparison of hysteresis losses in various samples of iron. Only a few years ago, a loss per lb. of 0·38 watts, under the above conditions, was not considered excessive for an average quality of commercial transformer iron : but great improvements have recently been made in the production of iron suitable for use in alternate current apparatus. The curve fig. 21 applies to a sample of iron giving a loss of 0·25 watts at $B = 2,500$, and an exceptionally good specimen tested by Professor Ewing in 1895 had a loss as low as 0·16 watts.

TABLE GIVING TOTAL LOSS IN WATTS PER POUND OF GOOD QUALITY, WELL INSULATED TRANSFORMER IRON STAMPINGS, ·014 INCH THICK

B., in C.G.S. lines per sq. in.	Frequency = 50			Frequency = 100		
	Hysteresis	Eddy Currents	Total	Hysteresis	Eddy Currents	Total
8,000	·036	·005	·041	·072	·019	·091
9,000	·041	·006	·047	·082	·024	·106
10,000	·047	·007	·054	·094	·030	·124
11,000	·054	·009	·063	·108	·036	·144
12,000	·061	·011	·072	·122	·043	·165
13,000	·069	·013	·082	·138	·050	·188
14,000	·077	·015	·092	·154	·058	·212
15,000	·086	·017	·103	·172	·067	·239
16,000	·095	·019	·114	·190	·076	·266
17,000	·102	·021	·123	·204	·086	·290
18,000	·111	·024	·135	·222	·096	·318
19,000	·122	·027	·149	·244	·107	·351
20,000	·135	·030	·165	·270	·118	·388
22,000	·150	·036	·186	·300	·143	·443
24,000	·175	·043	·218	·350	·170	·520
26,000	·200	·050	·250	·400	·200	·600
28,000	·222	·058	·280	·444	·232	·676
30,000	·248	·067	·315	·496	·266	·762
35,000	·312	·091	·403	·624	·363	·987
40,000	·382	·118	·500	·764	·473	1·237
45,000	·460	·150	·610	·920	·600	1·520

In the table on p. 72 the hysteresis losses have been taken from the curve fig. 21, and the eddy current losses have been calculated by means of formula (18).

It is interesting to note that the eddy current losses are of more relative importance, both for the higher frequencies and the higher inductions. With low frequency and induction the eddy current losses are almost negligible in well-laminated iron. Thus, for $B_{,,} = 15,000$ and $n = 50$ the eddy current loss in plates 0.014 inch thick is only equal to 16 per cent. of the total loss. On the other hand, if an exceptionally good quality of iron is used, thus permitting the use of higher inductions, the eddy currents at a frequency of 100 may form a considerable part of the total losses. For this reason tests made on transformers have sometimes shown the eddy current loss under working conditions to be fully equal to half the hysteresis loss.

27. **Vector Diagram for Inductive Circuit containing Iron.**—In § 23 (p. 57) it was shown that the power absorbed by eddy currents in the iron was supplied to the circuit in the form of a component of the total current practically opposite in phase to the induced volts, i.e. in phase with that component of the impressed volts which balances the induced volts; let us call this current component Cs.

In § 25 it was shown that the remaining portion of

the total current, i.e. that portion of it which is available for producing the magnetic induction, may be divided into two components—the 'wattless' component c_E, in phase with the magnetisation wave, and the 'hysteresis'

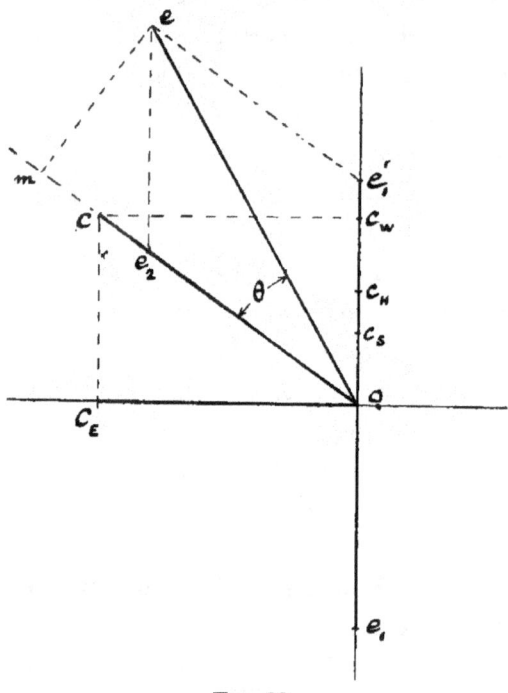

Fig. 22

component c_H, exactly one quarter period in advance of c_E, and therefore in phase with the eddy current component c_S.

In fig. 22 let oe_1 represent the induced E.M.F. Then c_E, the wattless component of the exciting current, will have to be drawn 90° in advance of oe_1. Draw c_S and c_H (both opposite to e_1) and add them together, to form c_W, the total 'work' component of the current. Now add the vectors c_E and c_W in the usual way, and draw oc, which will represent the total current.

If $R=$ the ohmic resistance of the circuit, e_2, the effective volts (in phase with c), will be equal to $c \times R$. Add e_2 and the vector e'_1 (exactly equal and opposite to e_1), and draw oe, which will represent the necessary impressed potential difference in order that the current c may flow through the circuit.

It is interesting to compare fig. 22 with fig. 12 (p. 53). The current in fig. 12 (which applies to a circuit or choking coil without an iron or metal core) lags exactly 90° behind e'_1; whereas in fig. 22 the current is no more 'wattless' with respect to this component of the total E.M.F. Also, an inspection of fig. 22 will show that the true watts supplied to the circuit are still equal to the product $c \times e \cos \theta$, and that the cosine of the angle θ may therefore still be defined as the ratio of the true watts to the apparent watts, or the '*power factor*' of the circuit.

In order to make this quite clear, the vector oe has been projected upon oc. Then

$$oc \times oe \ \cos \theta = oc \times om$$
$$= oc \times oe_2 + oc \times e_2m$$
$$= C^2R \text{ losses} + oc \times e_2m$$

$$\text{But } oc \times e_2m = oc \times \text{the projection of } oe'_1 \text{ on } oc$$
$$= oe'_1 \times \text{the projection of } oc \text{ on } oe'_1$$
$$= oe'_1 \times oc_{\mathrm{W}}$$
$$= oe'_1 \times oc_8 + oe'_1 \times oc_{\mathrm{H}}$$
$$= \left\{ \begin{array}{l} \text{watts lost by} \\ \text{eddy currents} \end{array} \right\} + \left\{ \begin{array}{l} \text{watts lost by} \\ \text{hysteresis} \end{array} \right.$$

The product $c \times e \; \cos \theta$ is therefore equal to the total watts lost, both in the copper of the circuit and the iron of the core. It follows also that the three voltmeter method of measuring the power supplied to an inductive circuit, which was described in § 22 (p. 55), may still be used when a portion, or even the greater part, of the power supplied is transformed into heat in the iron core.

28. **Design of Choking Coils.**—Although we have gone at some length into the question of the losses which occur when iron is introduced into an alternating current circuit, it must not be supposed that, in order to design a simple choking coil—such as might be used, for instance, on an arc lamp circuit—it is necessary to predetermine, with any degree of accuracy, the losses which will occur in the iron core. We have only to be careful that the induction in the iron is reasonably low in order that the temperature rise may not be excessive; the actual losses will then be practically negligible.

The above remarks do not, however, apply to transformers, in which the iron losses are of the greatest importance; they must be kept very small in order

that the efficiency of the transformers at light loads
may be as high as possible.

Let us suppose it is required to design a choking
coil for use in series with a 10 ampere arc lamp on a
100 volt alternating current supply; the object of the
choking coil being to reduce the voltage at the lamp
terminals to 40.

Since the E.M.F. of self-induction is very nearly 90°
out of phase with the current, and therefore with the
effective E.M.F. (for we are assuming the iron losses to
be negligible), the back E.M.F. to be produced by the
choking coil must not be $(100-40)=60$ volts, but
more nearly $\sqrt{100^2-40^2}=92$ volts.

This, therefore, is the voltage which would be
measured across the terminals of the choking coil, on
the assumption that the ohmic resistance of the winding
is also negligible.

In order to produce this back E.M.F., it is necessary
that the fundamental formula $e_m=\dfrac{4\mathrm{N}Sn}{10^8}$ (see p. 37)
should be satisfied. But this gives the relation between
the induction and the *mean* value of the induced volts.
We must therefore multiply e_m by the ' wave constant '
in order to get the $\sqrt{\text{mean square}}$ value of the induced
volts. Hence, $e=m\dfrac{4\mathrm{N}Sn}{10^8}$, and, on the assumption
that the induced E.M.F. follows the sine law—which
in nearly all cases will be sufficiently near to the truth

for our purpose—$m = 1 \cdot 11$, which enables us to write:

$$e = 92 = \frac{4 \cdot 44 \mathsf{N} S n}{10^8}.$$

Knowing the frequency n, we can therefore calculate the product $\mathsf{N}S$. It will be evident that we can vary the two factors, N and S (the number of turns in the coil) in whatever way we like, provided their product remains constant; in other words, S must vary inversely as N. Again, *for any given arrangement of the magnetic circuit* there is only one definite value of S which will give the required result, since the current passing through the coil is of constant strength (in this case 10 amperes). It is therefore necessary to know what value of the induction, or of the total flux N, corresponds to any given value of the magnetising ampere-turns. This relation may be determined either by experiment, or, if no great accuracy is required, by calculation in the usual way from measurements of the (proposed) magnetic circuit, and with the aid of the usual magnetisation or permeability curves.

It will almost certainly be found that a closed magnetic circuit is not suitable in the case we have taken as an example, because the number of turns, S, would have to be very small in order to keep the induction in the iron within reasonable limits, thus making it necessary for N—and therefore the cross-section of the core—to be very great. However, by introducing an air gap in the magnetic circuit, the resistance of the latter may be readily increased so as to allow of a proper number of turns being wound on the core.

In calculating the ampere-turns, it will generally be sufficiently accurate to assume that if we multiply the current by 1·4 we shall obtain its *maximum* value (in this case 14 amperes), which, as far as magnetic effects are concerned, is what we require to know.

The following empirical formula may also be of use: it gives the ampere-turns required per inch length of the iron circuit for a good quality of transformer iron:

$$\text{Ampere turns per inch} = \frac{B_{,,}}{10,000} + 2 \quad . \quad . \quad (22)$$

where $B_{,,}$ = the induction in the iron in C.G.S. lines per square inch of cross-section.

The above rule applies only to cases where $B_{,,}$ lies somewhere between 12,000 and 40,000, and it is based on the assumption that the curve connecting exciting force and induction is a straight line between these limits.

For air, the formula:

$$\text{Ampere-turns per inch} = 0{\cdot}313 B_{,,} \qquad . \quad (23)$$

is true for all values of $B_{,,}$.

CAPACITY

29. **Definition of Quantity and Capacity.**—
The unit quantity of electricity is the coulomb. A
coulomb is defined as the quantity of electricity conveyed
in one second when the current is one ampere.

The *capacity* of a condenser is the number of coulombs
required to be given to one set of plates in order to
produce a difference of potential of one volt between the
two sets of plates ; or, since the capacity of a condenser
is constant per volt difference of potential at the
terminals, whatever this difference of potential may be,
it may be defined as the ratio of the charge, in
coulombs, to the potential difference, in volts, between
the coatings.

Hence, if $K=$ capacity of condenser in farads,

$$V = \text{volts applied at terminals,}$$

the charge in coulombs $= K \times V$.

The *microfarad* is the one-millionth of a farad.

To calculate capacity ; let $K_m =$ the capacity in microfarads ; $A =$ the area of each set of plates in square

inches; $t=$ the distance between the plates; then, if the plates are separated by air,

$$K_m = \frac{A}{4\cdot452 \times 10^6 \times t} \qquad . \qquad . \qquad . \quad (24)$$

If the plates are not separated by air, the capacity will be found by multiplying the above value of K by the specific inductive capacity of the dielectric. This multiplier is about 3 for vulcanized rubber, 5 for mica, and 10 for very dense flint glass.

Capacity of Cylindrical Condenser.—If $l=$ the length in centimetres, D and d the diameters of the outer and inner coatings respectively, then

$$K_a = \frac{l}{2 \log_e \dfrac{D}{d}}$$

where K_a is the capacity in absolute *electrostatic* units, on the assumption that the specific inductive capacity is unity.

Since the microfarad is approximately 900,000 times greater than the absolute electrostatic unit, the above expression must be divided by this number in order to convert it into practical units, and if we put l_1 for the length in *feet*, and convert the Neperian logs. into common logs., we may write :

$$K_m = \frac{7\cdot353}{10^6} \times \frac{l_1}{\log_{10} \dfrac{D}{d}} \cdot \qquad . \qquad (25)$$

G

The following measurements of the capacity per mile of high-tension electric light cables have been kindly furnished by the makers; they all refer to 19/18 concentric cables, the capacity given being that between the inner and outer conductors :

British Insulated Wire Co. (Paper) . ·31 microfarads
W. T. Glover & Co. (Vulcanized Rubber) ·615 „
 „ „ (Diatrine) . . ·315 „

Condensers Connected in Parallel.—The joint capacity of a set of condensers connected in parallel is evidently equal to the sum of the several capacities.

Condensers Connected in Series.—The capacity of several condensers connected in series will be less than that of any single one of the condensers, because the arrangement is evidently equivalent to increasing the distance between the plates of any one condenser, and the reciprocal of the total capacity will be equal to the sum of the reciprocals of the several capacities, or

$$K = \frac{1}{\dfrac{1}{k_1} + \dfrac{1}{k_2} + \dfrac{1}{k_3} + \&\text{c.}}$$

A charged condenser must be considered as containing a store of electric energy which may be used for doing useful work by joining the terminals of the condenser through an electric circuit. In this respect it may with advantage be compared with a deflected spring, which, so long as it is kept in a state of strain, has a certain capacity for doing work.

30. **Current Flow in Circuit having appreciable Capacity.**—Consider an alternator (fig. 23) connected through the circuit R to the condenser K.   If we neglect the resistance of the circuit, the amount of current passing will evidently be determined by the E.M.F. e, and the capacity K of the condenser.

As long as the current is flowing in a positive direction, the condenser is being charged; when the current reverses (second half of complete period), the condenser is discharging.   It follows that the

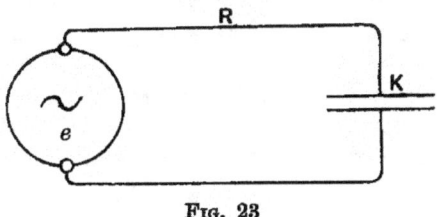

FIG. 23

maximum charge—and therefore the maximum value of the condenser E.M.F.—will always occur at the instant when the current is changing its direction. This is clearly shown in fig. 24.   Here the curve c represents the current flowing through the condenser. At the instant b, when, from having been flowing in a *positive* direction, it is passing through zero value before flowing out of the condenser again, the charge in coulombs (represented by the curve q) has evidently reached its maximum *positive* value d, and will now fall, through zero, to its maximum *negative* value, which

G 2

will occur at the moment when the current is changing from its negative to its positive direction—that is to say, at the moment when the *positive* current has ceased to flow into the *opposite set of condenser plates*.

When a condenser is charged by means of a current flowing in a *positive* direction, it is evident that the condenser E.M.F. will be *negative*—in other words, it will oppose the applied E.M.F. required to force the current into the condenser, and will therefore tend to expel the charge. Hence we may draw the curve e_1, exactly opposite to q, and such that its ordinates are at every point equal to $\dfrac{q}{K}$, where K is the capacity of the condenser. Since we are neglecting the resistance of the alternator and leads, the curve e_1, which is that of the potential difference at the condenser terminals, will be exactly equal and opposite to the alternator E.M.F. (fig. 23). Thus we see that the *condenser* E.M.F. is exactly one quarter period in *advance* of the current; whereas, when dealing with the question of self-induction, we found that the back E.M.F. of self-induction *lagged* exactly one quarter period *behind* the current.

31. Determination of Condenser Current.— Let K be the capacity, in farads, of a condenser connected in series with an alternating current circuit; and let V stand for the maximum value of the volts

at condenser terminals; then the total charge in
coulombs at the end of each half-period will be
equal to $K \times V$.

But *quantity = current × time*, hence maximum
charge in coulombs$= i_m \times \dfrac{1}{4n}$, where $i_m =$ the mean value

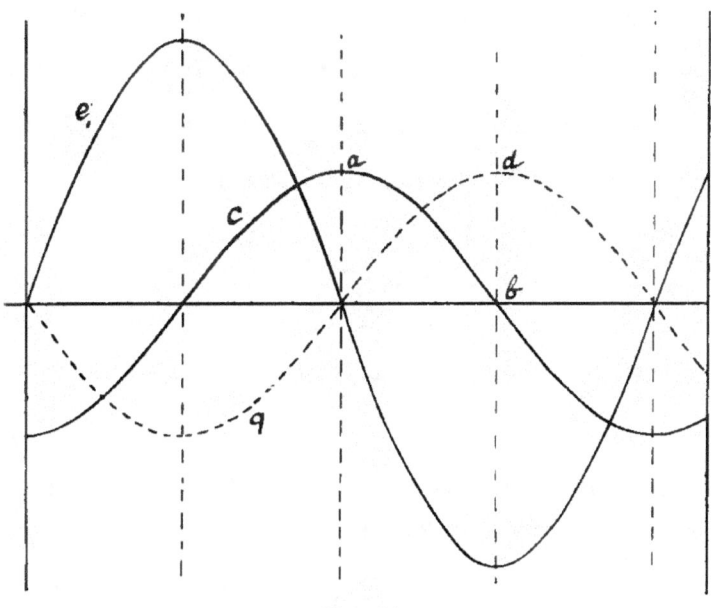

FIG. 24

of the current between zero and its maximum (i.e. from
a to b, fig. 24) and $\dfrac{1}{4n} =$ the time taken by the current
to change from its maximum to its zero value, n
being the frequency in complete periods per second.

It follows that $\dfrac{i_m}{4n} = KV$ or

$$i_m = 4nKV \qquad . \qquad . \qquad . \qquad . \quad (26)$$

The quantity we generally require to know being the $\sqrt{\text{mean square}}$ value of the current, let us write (see p. 37) $\dfrac{i}{i_m} = m$, where i stands for the $\sqrt{\text{mean square}}$ value of the current flowing through the condenser ; and $\dfrac{V}{v} = r$, where v is the $\sqrt{\text{mean square}}$ value of the condenser E.M.F. in volts. It follows that equation (26) may be written in the form :

$$i = (4mrn)Kv$$
$$= pKv$$

or, if the capacity is expressed in microfarads,

$$i = pK_m v \times 10^{-6} \qquad . \qquad . \qquad (27)$$

If the current is a sine curve, the multiplier p is equal to $2\pi n$, as already shown (§ 15), and

$$i = 2\pi nK_m v \times 10^{-6} . \qquad . \qquad (28)$$

32. **Vector Diagram for Current Flow in Circuit having appreciable Capacity.**—In fig. 25, oc is the current; oe_2, the effective component of the impressed potential difference (equal to $C \times R$, where R is the resistance of the circuit connected to the alternator terminals, see fig. 23); oe_1, the condenser

E.M.F., drawn at right angles to oc in the *forward* direction; and $o\grave{e}$ is then the necessary impressed potential difference, obtained by adding the force oe_2 to the force oe'_1 (exactly equal and opposite to e_1) in the usual way. The angle θ is the phase difference between the impressed potential difference and the resulting current, and it is a measure of the amount by which

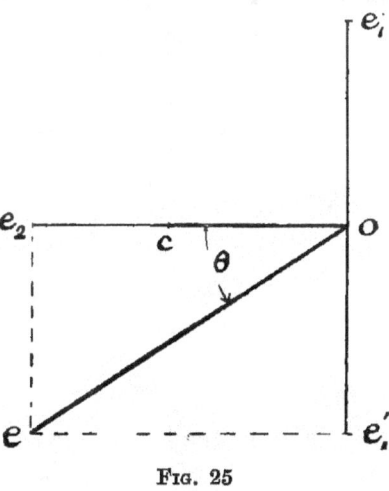

Fig. 25

the latter is in *advance* of the former. This diagram should be compared with fig. 12, p. 53.

Condenser in Parallel.—A problem of more practical utility than the above is that of a condenser or condensers connected in parallel with the main circuit; because this arrangement is almost exactly equivalent to the case of a concentric feeder, as used in nearly all systems of alternate current supply in this country. In fact, for all calculations of capacity current, &c. in long concentric mains, the concentric conductor may be considered as consisting of two ordinary conductors bridged across by a number of condensers.

In fig. 26 the alternator, which generates e volts at its terminals, is supplying a circuit of resistance $r + R$, and which, for simplicity, we will suppose to be without appreciable self-induction. This circuit is bridged by

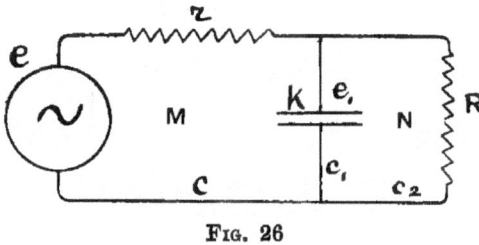

Fig. 26

the condenser K, the connections to which may be assumed to have no appreciable resistance.

Let c_2 be the current in the remote section, N ; it is evidently determined by the voltage e_1 at the condenser terminals, and, since we are assuming no self-induction in the circuit, it will be in phase with e_1 and equal to $\frac{e_1}{R}$. Also, the condenser current c_1 is a quarter period in advance of e_1 (for we are considering e_1 as the volts *supplied* to the condenser terminals) ; hence the total current, c, in section M, which is equal to the sum of these two currents, will be represented by the expression $\sqrt{c^2_1 + c^2_2}$.

In addition to the above, we have the relation $e_r = c \times r$, where e_r stands for that component of the applied E.M.F. which sends the total current c through the resistance r.

We are now in a position to draw the vector diagram for this particular circuit. Let us suppose the E.M.F. at condenser terminals (e_1) to be known. Draw

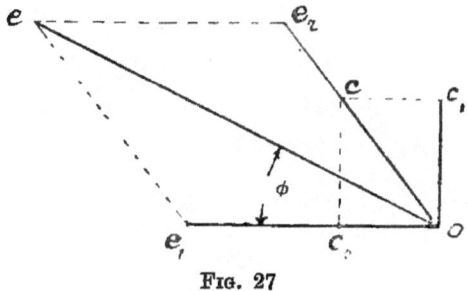

Fig. 27

oe_1 (fig. 27) to represent this force. Divide e_1 by R to get the current c_2 in the remote section, and lay down oc_2 to scale. Now calculate the condenser current $c_1 = pKe_1$ and draw oc_1 at right angles to oe_1 in the *forward* (clockwise) direction. The resultant oc of c_2 and c_1 is evidently a measure of the total current c in the section M. By multiplying c by the resistance r we obtain the effective E.M.F. e_r, required to overcome the resistance of the section M. It is now merely necessary to compound the two forces e_1 and e_r in order to obtain

e, the potential difference required at the alternator terminals in order that the voltage at the condenser terminals shall be e_1.

The angle ϕ, which represents the phase difference between the impressed volts on the whole system and the volts at condenser terminals, can be shown to be such that its trigonometrical tangent is equal to $\dfrac{pKRr}{R+r}$.

33. **Capacity Current in Concentric Cables.**—In very long concentric feeders, such as are used in many alternating current electric lighting systems, the current flowing into the cable when there is no connection at the distant end is sometimes quite appreciable, and, especially when the cables are insulated with vulcanized rubber and the frequency is high, it may amount to as much as 10 amperes, or even more, on a 2,000-volt circuit. This is the capacity current; and it follows from what has been said that if the load at the far end of the feeder be non-inductive, the in-going current at the station end will necessarily be greater than the out-going, or *work*, current.

If c=the in-going current,
c_2=the out-going current=60 amperes,
and c_1=the condenser current=10 amperes,
then $c=\sqrt{c^2_2 + c^2_1}=60.82$ amperes.

With regard to the statement that the current due to capacity is in *advance* of the impressed E.M.F., this is sometimes objected to on the ground that the effect cannot precede the cause, and it has even been suggested that we should speak of the current, not as *leading* by one quarter period, but as *lagging* by an amount equal to three quarters of a complete period. There is certainly no great objection to doing so, seeing that we are dealing with two periodically varying quantities, in connection with which there is no beginning and no end to be considered; but it is to be feared that such a way of stating the case would lead to some confusion regarding the manner in which an alternating current flows into and out of a condenser (see p. 83). It is not necessary to spend much thought on the question in order to see that a periodically varying current may well be in *advance* of a periodically varying E.M.F., without such a state of things being in any way equivalent to a flow of current occurring in a circuit *before* the existence of a difference of potential between its terminals; and, in any case, since it is a fact that the condenser current *is* in advance of the impressed alternating E.M.F., discussions as to whether or not it is correct to say so cannot rightly be considered as serving any useful purpose.

Before leaving this section on Capacity, in order to deal with a few general problems concerning the flow of currents in circuits possessing self-induction in addition to capacity and ohmic resistance, it should be stated that if this question of capacity has been treated less fully than the subject of self-induction, it is mainly due to the fact that the latter quantity is of very much more relative importance than capacity in alternating current work.

With regard to what has been called 'dielectric hysteresis,' and the other and more important causes of losses in condensers when used in connection with alternating currents, it is hardly necessary to concern ourselves with these; but it may be stated that these losses, in condensers with solid dielectrics, are approximately proportional to the square of the applied volts at the terminals. In fact, a commercial condenser behaves almost exactly as if a small non-inductive resistance (which may be termed the *spurious* resistance of the condenser) were joined in series with a theoretically perfect condenser, i.e. one in which there is no loss of energy when connected to a source of alternating E.M.F. It has also been found that the losses in a condenser are less after the latter has been thoroughly dried.

SELF-INDUCTION AND CAPACITY

34. Resistance, Choking Coil, and Condenser in Series.—The arrangement shown in fig. 28 consists of an alternator generating e volts at its terminals, and causing a current c to flow through a circuit of resistance$=r$; inductance$=pL$; and *reactance* (due to the condenser)$=\dfrac{1}{p\text{K}}$.

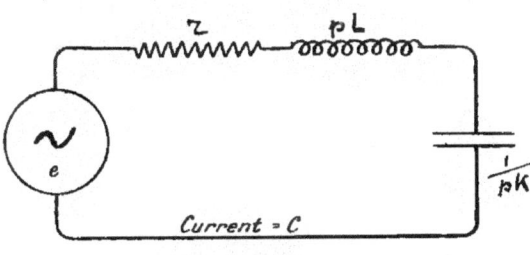

Fig. 28

The vector diagram, which gives the relation between the various component E.M.F.s and the current, is drawn in fig. 29.

Here oc represents the current, and oe_r the resultant or effective E.M.F., equal to $c \times r$, and in phase with the current; oe_L, at right angles to oc in the direction of retardation, is drawn to the same scale as e_r, and represents the total E.M.F. of self-induction; oe_K, also at right angles to oc, but in *advance*, is the condenser E.M.F. If the latter were equal to zero—which would be the case if the condenser were of infinite capacity, or if it

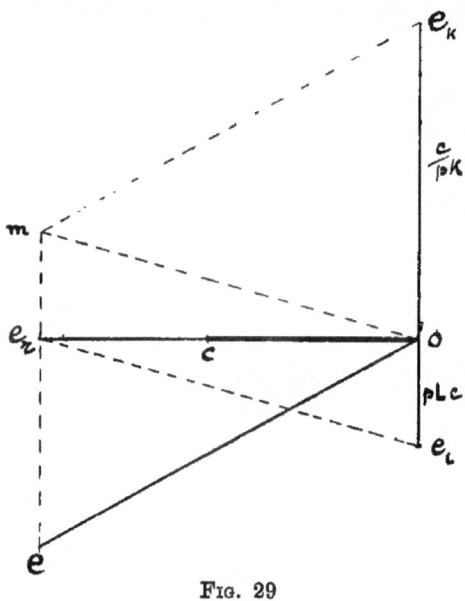

Fig. 29

were entirely removed—the total volts required at the alternator terminals would be equal to om, obtained by adding to e_r an E.M.F. exactly equal and opposite to e_L. But since the condenser E.M.F. is equal to oe_K, the volts actually required at the alternator terminals will be equal to oe, obtained by compounding om and a force exactly equal and opposite to e_K. Thus we clearly see how the effect of capacity is to counteract the effect of self-induction. The alternator E.M.F., instead of being in advance of the current by the amount indicated by the angle moe_r, now lags behind the current, as shown by the angle eoe_r. Also, since the distance e_re is equal to the difference between the condenser E.M.F. and the E.M.F. of self-induction, it follows that when these two forces are equal they will balance each other; the alternator volts will be equal to oe_r, and the circuit will behave in all respects as if it possessed ohmic resistance only. It is hardly necessary to point out that the slightest change in the frequency, or the capacity, or the coefficient of self-induction would instantly upset this balance. This is more especially the case with regard to the frequency, since the ratio of e_L to e_K is dependent upon the square of the frequency.

It will be evident from a glance at the diagram that either the condenser E.M.F. or the choking coil E.M.F. may be *greater* than the impressed E.M.F. applied to the circuit at the alternator end. That is to say, if we connect a voltmeter across the terminals of the condenser, or of the choking coil, the reading on this voltmeter may be higher than that which would be read on a voltmeter connected across the alternator terminals.

Let us, for simplicity, suppose that r, L, and n (the frequency) remain constant, but that the capacity, K, of the condenser can be varied at will; and let us plot a curve showing the relation between the capacity, κ, and the volts across the condenser terminals, on the assumption that the alternator volts remain constant.

This is easily done by merely altering the length oe_κ in the diagram, fig. 29, and calculating the ratio $\dfrac{e_\kappa}{e}$ for various values of κ.

An inspection of the diagram will, however, lead to one or two interesting conclusions.

Firstly,

$$\frac{oe_\kappa}{oe} = \frac{me}{oe} = \frac{\sin moe}{\sin ome}$$

but since r, L and n are assumed to be constant, the angle *ome* will remain unaltered, however much we may vary κ. Hence

$$\frac{e_\kappa}{e} \propto \sin moe,$$

and this ratio will, therefore, be a maximum when *moe* is a right angle.

Secondly, if r is expressed in ohms, κ in farads, and L (the coefficient of self-induction) in ' practical' units—i.e. in henrys—the condenser volts will be the same as the generator volts, or $oe_\kappa = oe$, when

$$\kappa = \frac{2L}{r^2 + p^2 L^2}.$$

This condition, it is evident, will also be more and

more nearly fulfilled as κ is made smaller and smaller without limit; for the lines me and oc will then both become very great, and sensibly equal in length.

If κ is *greater* than $\dfrac{2L}{r^2 + p^2L^2}$, e_κ will be *less* than e.

If κ is *less* than this amount, e_κ will be *greater* than e.

When $\kappa = \dfrac{L}{r^2 + p^2L^2}$, the ratio $\dfrac{e_\kappa}{e}$ is a maximum, for this is the condition which makes moe equal to 90°.

Thirdly, for a given impressed E.M.F., the current flowing in a circuit of constant resistance will evidently be a maximum when $oe_\kappa = oe_L$, for the whole of the impressed volts e will then be available for producing the current. This condition is fulfilled when:

$$\frac{1}{p\kappa} = pL \text{ or } p = \frac{1}{\sqrt{\kappa L}}.$$

In fig. 30 the curve already referred to has been plotted. The dotted line is the curve obtained by assuming r to be relatively large and L small; whereas the full line curve—which shows more clearly what is known as the *resonance* effect, i.e. the considerable rise in pressure at the end of a circuit containing capacity and self-induction for certain values of the capacity—is the result of considerably reducing the ohmic resistance of the circuit.

It should be noted that both curves rise from the value 1, when $\kappa = 0$, to a maximum when $\cdot\kappa = 3$, and

H

that they will again pass through the value 1 when
$\kappa = 6$—i.e. *twice* the critical value which gave the
maximum ' resonance ' effect.

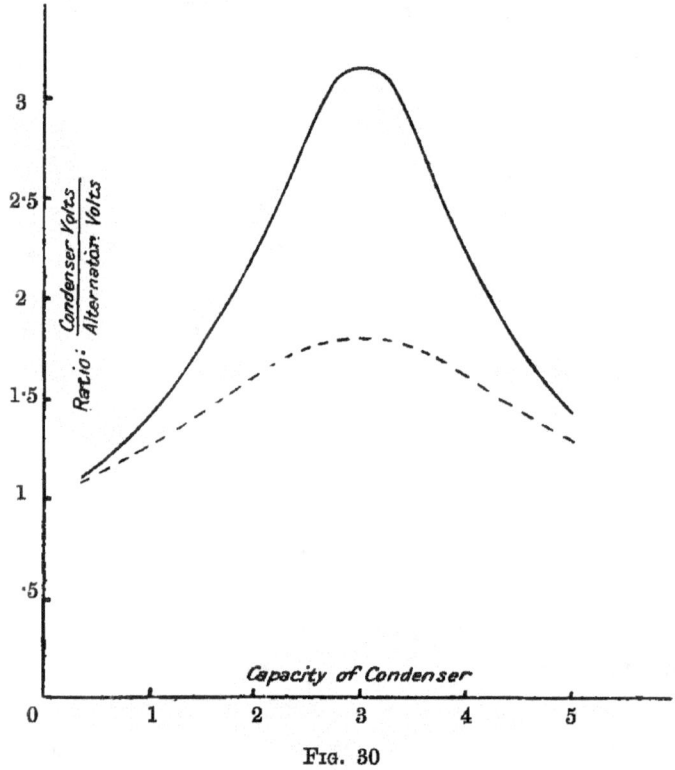

Fig. 30

Although we have considered the result of varying
the capacity only, it will not be necessary to dwell
further upon the behaviour of a circuit containing
capacity and self-induction in series.

The diagram, fig. 29, is so very simple that the effect of varying L while K is kept constant, or of keeping both L and K constant and varying the frequency, may be studied without much difficulty; and if curves are afterwards plotted in order to show the manner in which the volts (or the amperes) depend upon the variable quantity, these curves will be found to be of the same general appearance as those drawn in fig. 30.

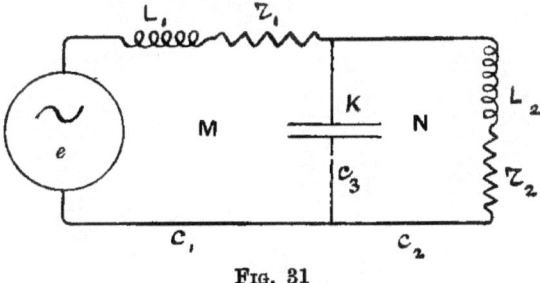

FIG. 31

35. Choking Coils and Resistance in Series; Condenser as Shunt.

In fig. 31 the circuit, which contains both resistance and self-induction, is shown bridged across by the condenser K.

Let c_2, r_2, and L_2 denote the current, resistance, and coefficient of self-induction of the remote, or condenser section, N; while c_1, r_1, and L_1 represent the value of these quantities in the alternator section M.

Assume the current c_2 to be known. Then, in the diagram, fig. 32, draw oe_2 $(=c_2 r_2)$ to represent the effective E.M.F. in section N. Erect a perpendicular at e_2, and make $e_2e_n = pL_2c_2$; then join oe_n. Evidently, since (see fig. 12, p. 53) oe_n is the impressed E.M.F. in the section N, it will be exactly equal to the condenser E.M.F., and we can therefore calculate the condenser current, $c_3 = pKe_n$. Now draw oc_3 at right angles with oe_n in the direction of *advance*, and

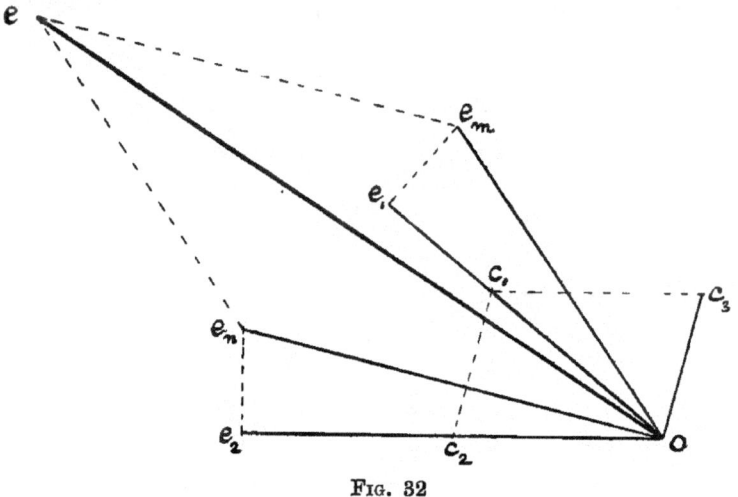

Fig. 32

add the current vectors oc_2 and oc_3 to obtain the total current c_1 in the section M. Proceed to calculate the necessary impressed volts in section M in order to produce this current c_1, on the assumption that the condenser is short-circuited. A similar construction to that used in connection with section N is applicable here; it is merely necessary to make $oe_1 = c_1r_1$, and

e_1e_m (at right angles to oc_1) $= pL_1c_1$. By compounding the vectors e_n and e_m in the usual way, the resultant oe is obtained, which is equal to the total impressed E.M.F., or the required potential difference at the alternator terminals.

It is interesting to note that the current c_2 in the remote section may very well be *greater* than the total current c_1 in the alternator section, and that the relation between the currents c_1 and c_2 depends upon the capacity of the condenser, and the frequency, and the amount of self-induction in the section N.

When $\kappa = \dfrac{2L_2}{r^2_2 + p^2L^2_2}$ these two currents will be equal.

If κ is *greater* than this amount, c_2 is *less* than c_1,

If κ is *less* than this amount, c_2 is *greater* than c_1.

Although many other combinations of circuits containing capacity and self-induction may occur in practice, it should not be necessary to consider any more special cases ; the method of constructing the vector diagrams, from which can be determined, by actual measurement, the values of the various alternating quantities, has been made, it is hoped, sufficiently clear to enable anyone to predetermine the relation between current flow and impressed E.M.F. in any circuit, even should the conditions differ somewhat from those in the problems already worked out. The fact that we have generally—for simplicity of construction—assumed to be known one of those quantities which it is the object of the diagram to determine, is of little consequence, seeing that the finished diagram gives the proper relations between the various E.M.F.s and currents, and that a simple proportion sum on the slide rule enables us to express the lengths scaled off the diagram in the proper units.

There is, perhaps, one more problem which may, with advantage, be considered before taking up the question of mutual induction, and that is the case of a number of circuits joined in parallel across the terminals of a constant pressure alternating current source of supply.

We will assume the various branches of the circuit to have ohmic resistance and self-induction only, in order that the diagrams may remain clear, and simple of construction; but it should be understood that the introduction of capacity in the circuits does not really increase the difficulties of the problem; and the exceedingly neat geometrical solution given below, and published by Mr. Alex. Russell in the ' Electrician ' of January 13, 1893, is equally applicable to the case of branched circuits containing resistance, capacity, and self-induction, jointly or otherwise, in any or all of the branches of the divided circuit.

36. **Current Flow in divided Circuit.**—Let r_1 , r_2 . . . be the resistances of the various branches ;

$L_1, L_2 \ldots$ the coefficients of self-induction, and $c_1, c_2 \ldots$ the currents. Then, if e is the potential difference between the points a and b (see fig. 33), the relation between the current c_1 in any one branch and the impressed E.M.F. e will be given by the expression :

$$e = c_1\sqrt{r_1^2 + p^2 L_1^2} \qquad . \qquad . \qquad . \quad (29)$$

where the quantity $\sqrt{r_1^2 + p^2 L_1^2}$ is what has been called the *impedance* (see p. 54).

It should, however, be pointed out that the above relation is only true on the assumption that there is no

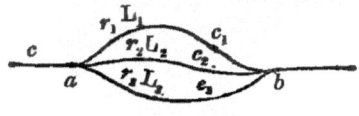

Fig. 33

'*mutual induction*' between the branches of the circuit ; that is to say, that the magnetism due to the current in any one branch does not induce any appreciable E.M.F. in any other branch.

In order that we may graphically determine the various currents and their phase differences, let us draw, in fig. 34, two lines, OX and OY, at right-angles to each other, and mark off on OX the distances $Or_1, Or_2 \ldots$ equal to $r_1, r_2 \ldots$; and $OL_1, OL_2 \ldots$ on OY equal to $pL_1, pL_2 \ldots$ Join $r_1L_1, r_2L_2 \ldots$ Then the length of the lines $r_1L_1, r_2L_2 \ldots$ is a measure of the impedances of the various branches, and the angles a_1, a_2 . .

represent the amount by which the currents lag behind
the impressed volts e.

Now draw a circle (fig. 35) of diameter oo_1 somewhat
greater than the longest of the impedance lines in
fig. 34, and mark off the chords oa_1, oa_2 . . . equal in
length to the lines r_1L_1, r_2L_2 . . . of fig. 34.   Join oa_1,
oa_2 . . . and produce to c_1, c_2 . . . where these lines

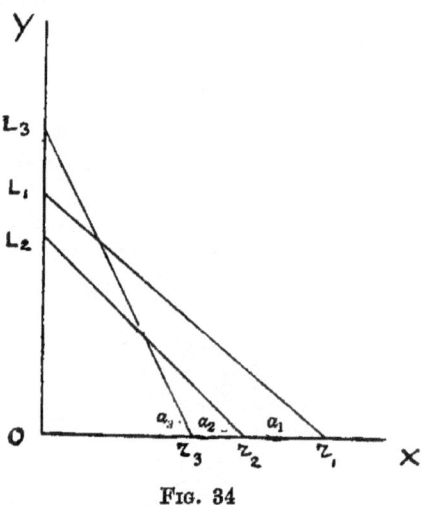

Fɪɢ. 34

cut the tangent o_1x, which will be at right angles to
the vertical diameter oo_1.

Now, because the dotted line o_1a_1 is at right angles
to the hypotenuse oc_1 of the triangle o c_1 o_1, it can easily
be shown that $oc_1 \times oa_1 = (oo_1)^2$, and this being true of all
the other triangles, we may write :

$$(oo_1)^2 = oc_1 \times oa_1 = oc_2 \times oa_2 = \&c.$$

Hence, since the line oa_1 represents the quantity $\sqrt{r_1{}^2 + p^2L_1{}^2}$ in equation (29), it follows that the

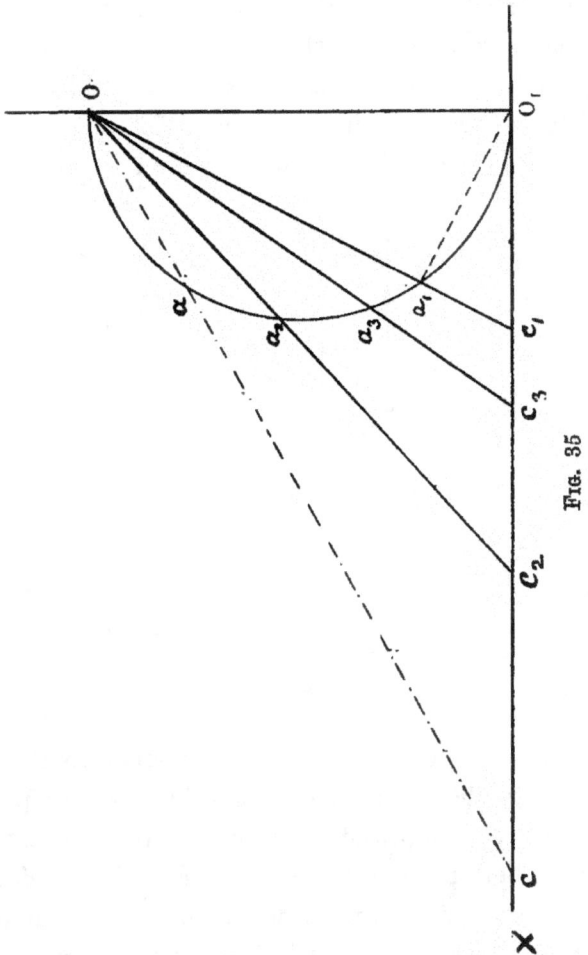

Fig. 35

lengths of the lines oc_1, oc_2 . . . are proportional to the currents in the different branches.

By drawing, in fig. 36, the line Oe to represent the
volts e, and the various lines Oc_1, Oc_2 . . . to represent
the currents in the branches—of which we now know
both the magnitude and the angles of lag—we can, by
compounding these current vectors, obtain the vector Oc,
which gives the magnitude of the total current c, and
the amount by which it lags behind the impressed
volts e.

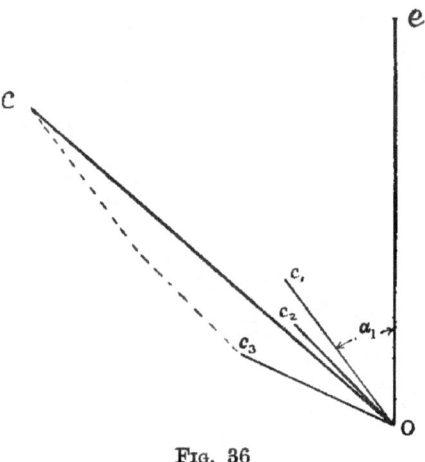

Fɪɢ. 36

By drawing Oc, in fig. 35, proportional to the total
current, we obtain the length Oa, which is that portion
of the line Oc comprised between the point o and the
point a where it cuts the circle; and this is evidently
equal to the joint impedance, or to that quantity by
which it is necessary to divide the volts e in order to
obtain the resulting current c.

In solving the above problem we have not considered that part of the complete circuit beyond the
points a and b—i.e. that portion of the circuit which
contains the source of supply.   This is, however, easily
taken into account, as it is merely necessary to consider
the divided portion of the circuit as a single conductor
of *impedance* oa (fig. 35) in series with the remaining
portion of the circuit, containing the alternator or other
source of supply.

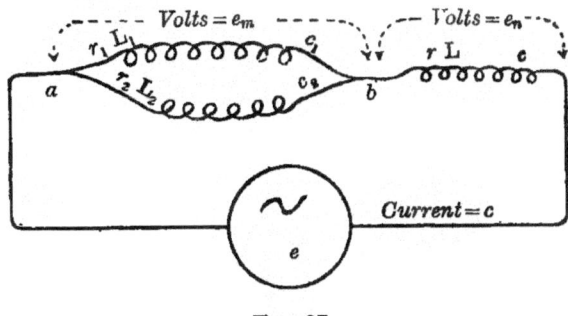

FIG. 37

Let us consider the arrangement shown in fig. 37.
Here r is the resistance and L the coefficient of self-
induction of that portion of the circuit containing the
alternator, and which we will suppose to consist of a
single conductor.

Instead of adopting the construction used in the
previous example, we may proceed in a slightly
different manner, and work out the problem by means
of a single diagram, as shown in fig. 38.

Here the vertical line oe_m represents the (assumed) pot. diff. between the points a and b, as shown by oe in fig. 36.

On oe_m, as a diameter, describe the dotted semi-circle as shown. Draw oe_{c1} in such a direction that the

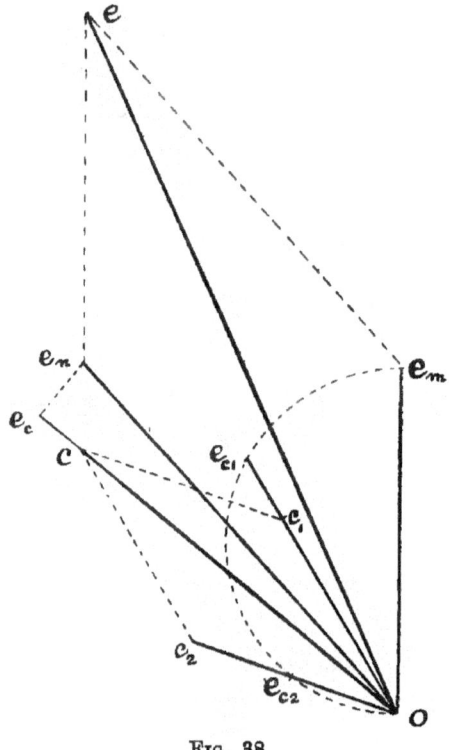

Fig. 38

tangent of the angle $e_m oe_{c1}$ is equal to $\dfrac{pL_1}{r_1}$. The line oe_{c1} will then represent in magnitude and phase the *effective* component of the E.M.F. in the branch r_1L_1.

By dividing this quantity by r_1 we obtain the current c_1 in this branch. The current c_2 in the other branch is obtained in a similar manner.

Now compound c_1 and c_2 to obtain the total current c, which, when multiplied by r, will give us e_c, the *effective* component of the total E.M.F. required to overcome the resistance r of the alternator portion of the circuit. At the point e_c erect the perpendicular $e_c e_n$, of which the length is proportional to the quantity pLc, or the E.M.F. of self-induction of the coil rL. Join Oe_n, which gives us the total E.M.F. required to send the current c through the undivided portion of the circuit. By compounding this force e_n with the force e_m required to overcome the impedance of the divided circuit, we obtain the total necessary impressed E.M.F. e for the circuit in question when the total current flowing is c amperes.

This diagram (fig. 38) is very instructive, although it is somewhat complicated in appearance. We will simplify it by assuming both L and L_1 to be negligible, and also r_2 (see fig. 37), which leaves us a non-inductive resistance, r, in series with a choking coil L_2, the latter being shunted by a non-inductive resistance r_1. As such an arrangement leads to some interesting results, it will be advisable to devote a separate paragraph to its consideration.

37. Shunted Choking Coil in series with Non-inductive Resistance.—In the diagram, fig. 39, let oc_2 be the current through the choking coil L_2 (fig. 37), of which the resistance is assumed to be negligible. Then oe_m, at right angles to oc_2 and equal to pL_2c_2, will be the potential difference between the

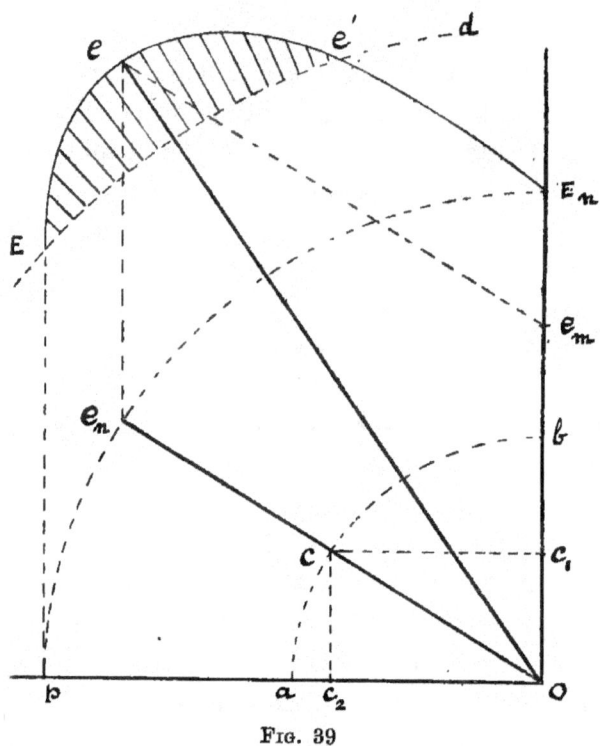

Fig. 39

points a and b. The current c_1 through the *non-inductive* resistance r_1, in parallel with the coil L_2, will be in phase with e_m, and equal to $\dfrac{e_m}{r_1}$. By adding the vectors oc_1 and oc_2 we obtain oc, the total current in the circuit, which, multiplied by the value of the *non-inductive* resistance r, gives us the volts e_n across the terminals of this resistance. The total necessary impressed E.M.F. e is obtained, as before, by adding together e_n and e_m in the usual way.

Let us now assume that the resistance r and the coefficient of self-induction L_2 remain constant, while we try various values of the shunt resistance r_1, and note the manner in which the total E.M.F. e will have to vary *in order that the total current c may remain constant.*

By making r_1 increase from 0 to ∞ —which may be done by starting with the choking coil short-circuited, and gradually increasing the resistance r_1 until the coil is unshunted—the point c (fig. 39) will move from b to a on the dotted circle described from the centre o. The point e_n—since $e_n = cr$—will move from E_n to p on a circle described from the same centre; and the point e —as will be found by constructing the diagram for various positions of the line oc—will follow the curve $E_n eE$; the point E_n representing the value of e (equal to e_n) when $r = 0$, whereas E (equal to $\sqrt{e^2_m + e^2_n}$) gives us the value of e when there is no shunt to the choking coil.

By describing through E the dotted circle Ed from the centre o, we see that, for a *constant total current* passing, the maximum value of the impressed volts e

does not occur when the potential difference between the points a and b (fig. 37) is at its maximum—or, in other words, when, by making $r = \infty$, the full effect of the choking coil is obtained—but it corresponds to that particular value of the shunt resistance r_1 which gave us the point e obtained in the first instance.

If, however, the resistance r_1 is reduced beyond the value which brings e to the point e', the total impressed E.M.F. will fall below what it was when the coil was unshunted.

The explanation of this effect is, of course, to be found in the fact that, although the addition of a non-inductive shunt to the choking coil reduces the current passing through the latter, and consequently also the back E.M.F. of self-induction, it brings the total current—and therefore the volts e_n across the resistance r —more nearly in phase with the impressed volts e. The result is that, of these two causes tending to produce opposite effects, the latter will, under certain conditions, overbalance the former ; and we thus see that the effect of shunting a choking coil which is connected in series with a non-inductive resistance is—for certain relative values of r_1, L_2 and r—to *increase* the total impedance of the circuit.   In other words, if we maintain a constant difference of potential e between the ends of the circuit, the effect of joining a non-inductive resistance in parallel with the choking coil may be to *reduce* the total current in the circuit ; a result which illustrates very clearly the fact that the back E.M.F. due to the choking coil is not in phase with the main E.M.F., and therefore is of far less use in choking back the current than would be an equal drop in volts due to the introduction of a non-inductive resistance.

38. Impedance of Large Conductors.— Imagine a straight length of cable, of fairly large cross-section, through which a steady continuous current is flowing. The lines of magnetic induction due to this current will not all pass through the non-conducting medium surrounding the conductor; but a certain number of them—due to the current in the central portions of the cable—will pass through the substance of the conductor itself; in other words, the magnetic flux surrounding one of the central strands of the cable will be greater than that which surrounds a strand of equal length situated near the surface. It follows that if the circuit be now broken, the current will die away more quickly near the surface of the conductor than at the centre; and, for the same reason, on again closing the circuit, the current will spread from the surface inwards.

If, now, we imagine the conductor to be used for conveying an alternating current, it is evident that, with a sufficiently high frequency (or even with a low frequency if the conductor be of large cross-section), the current will not have time to penetrate to the interior, but will reside chiefly near the surface of the cable,

I

This crowding of the current towards the outside portions of the conductor has the effect of apparently increasing its resistance; and it follows that, if c is the total current in the cable, and r its ohmic resistance, the power lost in watts will no longer be c^2r, as would be the case if the current were a steady one; but c^2r', where r', which stands for the apparent resistance of the cable, is greater than r, its true resistance.[1]

It is hardly necessary to remark, after what has already been said, that the shape of the cross-section of a heavy conductor plays an important part in determining what Mr. Perren Maycock has called the *conductor impedance*.[2] A conductor which presents a large surface in comparison with its sectional area will be the most economical for conveying alternating currents; and, for the same weight or cross-section,

[1] On account of the uneven distribution of the current density when the applied E.M.F. is a rapidly alternating one, the total self-induction of a given length of the cable, for the same maximum value of the current, will be slightly *greater* with a steady continuous current—in which case the current flow is evenly distributed over the whole cross-section—than with an alternating current, which does not utilise the central portions of the cable to the same extent; the result being that, although the flux of induction *outside* the bounding surface of the cable will be the same as before, the lines which pass through the space occupied by the material of the conductor are now fewer than in the first instance.

[2] The old term ' skin resistance ' was probably intended to remind one of the fact that the resistance of a conductor, when conveying a high frequency current, is apparently greater at the centre than near the surface or *skin*. To most people it would, however, suggest the contrary. ' Skin conductance ' would have been nearer to the mark, but almost equally inaccurate.

the 'drop' of E.M.F. with a given current will be less for a tube or flat strip than for a solid cylindrical or square conductor.

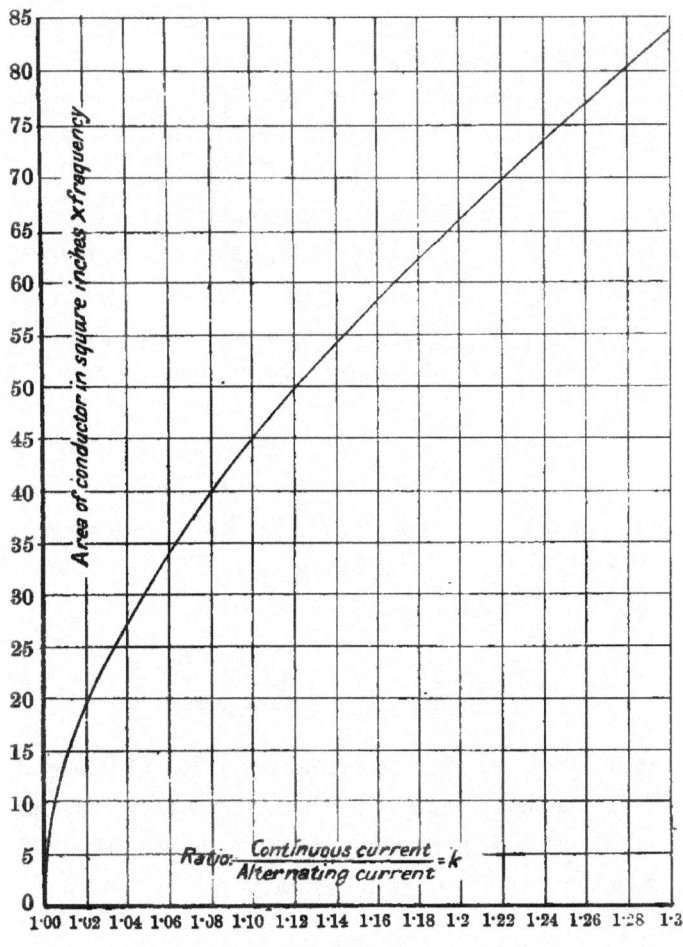

Fig. 40

By the aid of the curve (fig. 40), which has been drawn from data given by E. Hospitalier in ' L'Industrie Electrique,' the ' conductor impedance ' of cylindrical cables may be calculated for various diameters and frequencies.

The cross-sectional area of the cable in square inches multiplied by the frequency (in periods per second) gives us the point on the vertical scale from which, by means of the curve, the multiplier k may be read off the horizontal scale. If, now, we multiply the actual ohmic resistance of the conductor by k we obtain the impedance (see p. 54), or apparent resistance of the conductor when used for alternating currents of a given frequency.

The values of k obtained from the curve (fig. 40) are correct only if the conductor is of *copper*. To obtain k for any other ' non-magnetic ' cylindrical conductor, the product of the sectional area and frequency must be multiplied by the ratio

$$\frac{\text{conductivity of metal of conductor}}{\text{conductivity of copper}}$$

before obtaining the corresponding value of k from the curve. If the conductor is of iron or other ' magnetic ' material, the value of k may be very much greater than the number read off the diagram.

MUTUAL INDUCTION. TRANSFORMERS.

39.—In a single phase alternating current transformer there are two distinct circuits, the primary winding and the secondary winding, so arranged that the whole or a part of the magnetism due to a current in one of the coils passes also through the other coil.

The path of the magnetism is usually through a closed iron circuit; and, although in practice there is always a certain amount of leakage or stray magnetism which is not enclosed by the secondary coil, we will first, for the sake of simplicity, consider the case of an ideal transformer in which the whole of the magnetic flux set up in the primary passes also through the secondary coil.

40. **Theory of Transformer without Leakage.**—Consider a transformer of which the primary winding is connected to constant pressure mains, and from which no secondary current is taken. Under these conditions the primary circuit acts simply as a choking coil, of which the self-induction is so great and

the ohmic resistance relatively so small that no current
passes, except the very small amount required to
magnetise the core. The induced E.M.F. is therefore
practically equal and opposite to the applied potential
difference at primary terminals, and the relation
between the magnetic flux in the core, and the primary
impressed E.M.F. will therefore be given by equation (12) (see p. 37) or

$$e_m = \frac{4NSn}{10^8} \qquad . \qquad . \qquad . \quad (12),$$

where e_m now stands for the *mean* value, in volts, of
the primary E.M.F.

The number of turns, S, in the primary of a well-
designed transformer is always such that the current
required to produce the magnetisation N is very small ;
it is generally somewhere between 2 per cent. and 5 per
cent. of the full load current.

Although the rise and fall of the magnetism will be
$\frac{1}{4}$-period out of phase with the E.M.F., the open
circuit primary current will not be entirely ' wattless,'
but may be considered as made up of two components,
the ' wattless ' or true magnetising component, in phase
with the magnetism, and the ' work ' component, due to
hysteresis and eddy currents, in phase with the impressed E.M.F. The reader is, however, referred to
§ 27 (p. 73), where the question of magnetising current
for a circuit containing iron has already been dealt with.

Since the secondary and primary coils are both

wound on the same core, it follows that any variation in
the magnetism will produce an E.M.F. in *both* circuits ;
also, the actual volts induced will be directly proportional to the number of turns of wire in the coil.

On closing the secondary circuit through a resistance, the resulting current will produce a magnetising
force in the core. This magnetising force will not

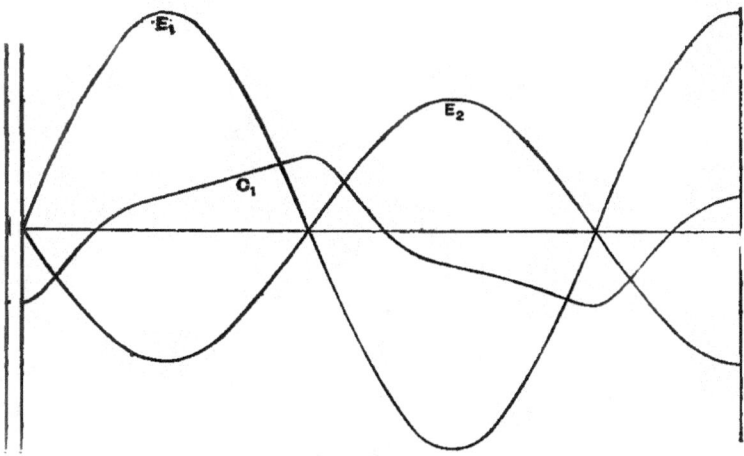

Fig. 41

produce a change in the magnetism, because it will
be instantly counteracted by a change in the primary
current, which will so adjust itself as to maintain the
same (or nearly the same) cycle of magnetisation as
before ; this being the only possible way by which
Ohm's law will continue to be fulfilled in the primary
circuit.

The curves (figs. 41 and 42) show the action of a transformer on open secondary circuit and on full load.

In fig. 41, E_1 is the curve of primary impressed E.M.F., and C_1 is the magnetising current, distorted as in fig. 19 (p. 65) by the hysteresis of the iron core. E_2 is the curve of secondary E.M.F., which coincides in phase with the primary induced E.M.F., and is therefore, on

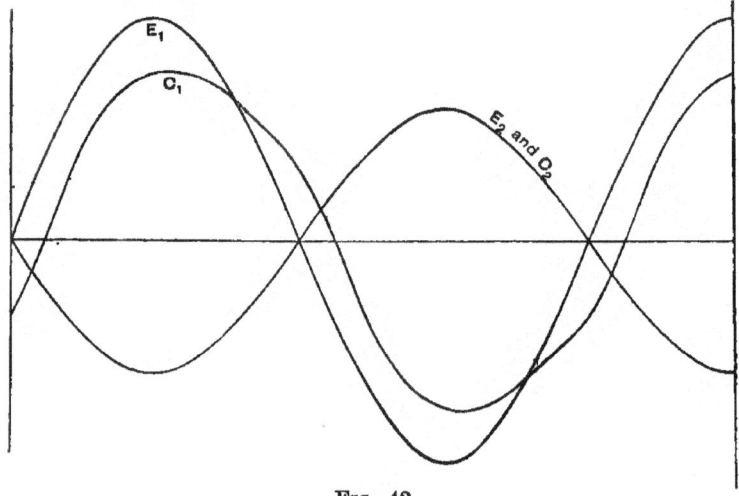

Fig. 42

account of the comparatively small ohmic resistance of the primary, almost exactly in opposition to the impressed E.M.F. The curve of magnetisation (not shown) would be exactly $\frac{1}{4}$ period in advance of the secondary E.M.F.

In fig. 42 the secondary circuit is closed through its proper load of incandescent lamps. There is no appreciable self-induction in such a circuit, and the secondary current will therefore be in step with the secondary E.M.F. For simplicity it is represented in fig. 42 by the same curve as E_2.

The tendency of this secondary current being to weaken the magnetism in the core, and therefore diminish the primary induced E.M.F., it follows that the current in the primary will grow until the magnetism is again of such an amount as to restore balance in the primary circuit. Hence, neglecting the small loss of volts due to the increased primary current, the induction through the primary must remain as before; and the new current curve, c_1 (fig. 42), is obtained by adding the ordinates of the current curve in fig. 41 to those of another curve, exactly opposite in phase to the secondary current, and of such a value as to produce an equal magnetic force.[1]

41. **Method of obtaining the Curve of Magnetisation from the E.M.F. Curve.**— In § 18 (p. 44) it was shown how, by differentiating the curve of magnetism, the curve of induced E.M.F. could

[1] In the above brief account of the action of a transformer no mention has been made of the constant coefficient of mutual induction (generally denoted by the letter M), which is frequently to be met with in the analytical treatment of the subject.

The reason of this omission is that all commercial transformers have iron cores, and the coefficient M can only be assumed to be constant if the transformer has an air—or other 'non-magnetic'—core.

It is, however, useful to know what is meant by the coefficient mutual induction, and, in order to understand this, it will be

be obtained.   It follows that if the E.M.F. curve is known, we have only to *integrate* it in order to obtain the curve of magnetisation.

The manner in which this may be done on drawing paper was fully explained by Dr. J. A. Fleming in his ' Cantor Lectures ' delivered in 1896.[1]

advisable, in the first place, to consider what is the *work done in moving a conductor carrying a current through a magnetic field.*

When Q units of electricity are moved against a potential difference $(V_1 - V_2)$, the work done is

$$w = Q \ (V_1 - V_2) \quad . \quad . \quad . \quad . \quad (a)$$

Also N lines of induction cut in the time t generate an E.M.F. $(V_1 - V_2) = \dfrac{N}{t}$.

Now, if i = the current through the conductor in amperes, Q (in absolute units) $= \dfrac{it}{10}$, and, by substituting this value of Q in equation (a), we get :

$$w \text{ (in ergs)} = \frac{N}{10} \quad . \quad . \quad . \quad . \quad (b).$$

Consider, now, two circuits, A and B, which we will suppose, for simplicity, to consist each of a single turn of wire. Suppose that a current i_a in A sends N_a magnetic lines through B, and that at the same time a current i_b in B sends N_b magnetic lines through A. If, now, the two circuits be moved apart until there is no *mutual induction* between them, a certain amount of work will be done ; and the work done in drawing N_a lines from B must obviously be equal to that done in withdrawing N_b lines from A. Hence by equation (b)

$$i_a \, N_b = i_b \, N_a \, ,$$

and if $i_a = i_b$, then $N_a = N_b$; and the constant coefficient of mutual induction, M, between two circuits may be defined as the total flux of induction through any one of them due to unit current in the other.

[1] As drawn by Dr. Fleming, the curve of induction *lags* $\frac{1}{4}$ period behind the induced E.M.F. curve, whereas it should, of course, precede

Let E in fig. 43 represent the curve of secondary,
or induced E.M.F. ; it will be practically of the same
shape as the curve of applied primary E.M.F., only of
opposite phase. At the point a where the curve crosses

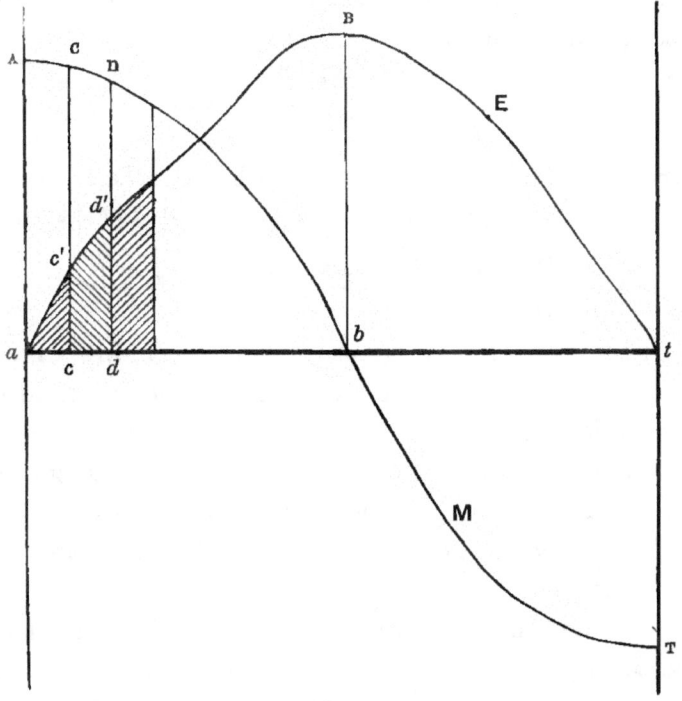

the datum line and E=0, the rate of change of the
induction to which the E.M.F. E is due will evidently

the latter by ¼ period. His diagram is, however, correct if the
E.M.F. curve be taken to represent the *impressed* terminal potential
difference.

also be zero; that is to say, the tangent to the curve of
induction M will be horizontal (see § 18, p. 44), and the
latter will therefore have reached its maximum value.
The length of this maximum ordinate, a_A, may be expressed in terms of the area of the E.M.F. curve a_Bt,
because, by formula (12) (p. 37),

$$e_m = \frac{4NSn}{10^8}$$

where $N =$ the total magnetic flux through the core,

$=$ the induction $B \times$ the cross-section of the core,

$S =$ the number of turns in the coil,

$n =$ the frequency,

and $e_m =$ the *mean* value of the induced E.M.F., which
is evidently equal to the area of the curve a_Bt divided
by the length at.

Having obtained the relation between the length of
the ordinate of the induction curve and the area of the
E.M.F. curve, we are now in a position to draw the
curve M.

Divide up the curve E into a number of small
portions (shown shaded) by means of the vertical lines
c_C, d_D, &c. Then the ordinate c_C of the curve M will be
less than the maximum ordinate a_A by an amount proportional to the area acc'. Similarly, the amount to
subtract from a_A to obtain the ordinate d_D will be
proportional to the area add', and so on. When half
the area of the semi-wave a_Bt has been reached, the

value of M will be zero; the ordinate b_B will therefore divide the curve aBt into two parts of equal area, as was previously pointed out in § 18 (p. 45). At t, where E is again equal to zero, the ordinate t_T of the curve M will be equal to a_A, but will be measured below the datum line, as the induction is now in a negative direction.

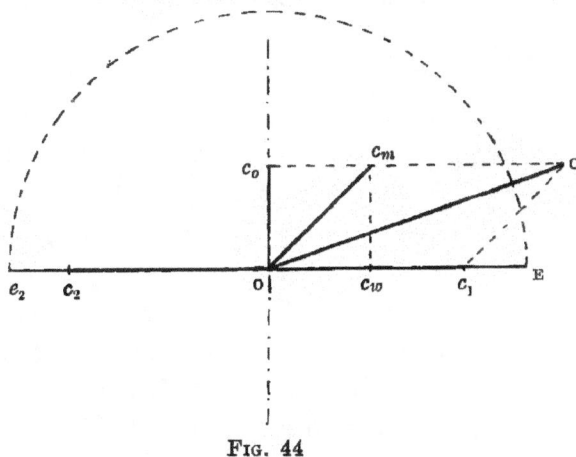

FIG. 44

42. **Vector Diagrams of Transformer without Leakage.**—Let us first consider a parallel, or constant pressure transformer [1] on open circuit.

In fig. 44 let oe_2 represent the secondary E.M.F. Then, on the assumption that the voltage drop in primary, due to ohmic resistance, is negligible—which assumption is always permissible—the primary im-

[1] As distinguished from a series, or constant current transformer.

pressed E.M.F., E, will be exactly opposite in phase to the secondary or induced E.M.F., and OE will be equal to Oe_2 multiplied by the ratio of turns in the two windings. In order to simplify the construction we shall here—and in all following transformer diagrams—assume the number of turns in the primary coil to be equal to the number of turns in the secondary coil. The transforming ratio will therefore be $\dfrac{1}{1}$ and the length OE will be equal to Oe_2.

With regard to the magnetising current in the primary coil, this, as already stated, will be very small; it will consist of the 'wattless' or true exciting current Oc_0 in phase with the induction—and therefore $\frac{1}{4}$ period in *advance* of Oe_2—and the 'work' component Oc_w in phase with OE. This 'work' component is due partly to hysteresis and partly to eddy currents in the iron core, as was shown in § 27 (fig. 22).

The total magnetising current Oc_m may now be drawn; it will lag behind the impressed E.M.F. by about 45°; the *power factor* (see p. 53) of a good closed iron circuit transformer, when no current is taken out of secondary, being between ·68 and ·74.

Effect of closing Secondary on Non-inductive Load.—The load being non-inductive, the secondary current Oc_2 (fig. 44) will be in phase with the secondary E.M.F. It will be balanced by a component, Oc_1, of the primary current, exactly equal and opposite to Oc_2 (and therefore in phase with E) ; and the total primary current will now be represented by the resultant oc.

Effect of closing Secondary on Partly Inductive Load. In fig. 45 let oe_2 be the secondary E.M.F. as before, and oc_2 the secondary current, which now lags somewhat behind this E.M.F. The balancing component of the primary current will still be equal and opposite to oc_2, with the result that the primary current oc will also

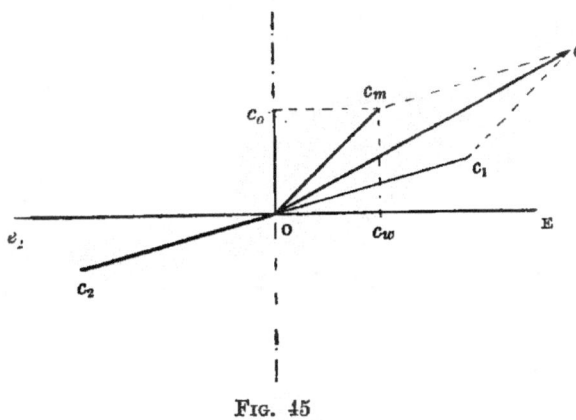

FIG. 45

lag behind the impressed E.M.F. It will be evident from inspection of the diagram that the energy put into the primary is still in excess of the energy taken out at the secondary terminals by the amount lost in hysteresis and eddy currents in the core.

Effect of Ohmic Drop on the Diagram.—So far we have not considered the effect of the resistance of the coils when the larger current is passing through the trans-

former. Taking this into account does not really add
to the difficulties of the problem, but it somewhat
complicates the diagrams.

In fig. 46 the complete diagram for a transformer
without magnetic leakage when working on a non-
inductive load (such as glow lamps) has been drawn.

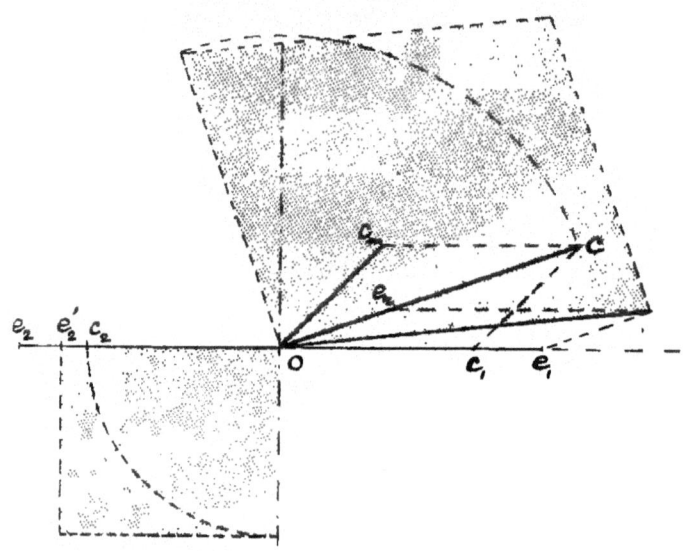

FIG. 46

Here e_2 and c_2 represent, as before, the secondary
E.M.F. and current. The voltage at the secondary
terminals will now no longer be e_2 but e'_2; the difference
between oe_2 and oe'_2 being equal to $c_2 \times r$, where r stands
for the resistance of the secondary winding.

The primary current oc will be obtained as before, but the necessary primary impressed E.M.F. will no longer be equal and opposite to e_2, as in fig. 44. It will be obtained by compounding oe_1 (exactly equal and opposite to oe_2) and oe_r, this latter quantity being the E.M.F. required to overcome the resistance of the primary coil; it must, of course, be plotted on the current vector oc, and its value is $c \times R$, where R stands for the resistance of the primary coil.

By constructing the shaded parallelograms as explained in § 20 (see fig. 12, p. 53), we have at once a graphical representation of the relation between the power supplied to the primary circuit and that which is taken out at the secondary terminals.

In an actual transformer the wattless component of the magnetising current is so small relatively to the total current c, that for all practical purposes the power supplied to the transformer when fully loaded on a non-inductive resistance is equal to $c \times$ E.

43. Transformers for Constant Current Circuits.—Since the primary and the secondary ampere-turns in a closed iron circuit transformer almost exactly balance each other—or, in other words, are equal and opposite—it follows that if the primaries of a number of transformers are all connected in series on a constant current circuit, the current through the secondary of each transformer will remain constant,

K

notwithstanding variations in the resistance of the secondary circuit.

Such an arrangement of transformers may therefore be used for arc lighting circuits, in cases where it is not desirable to transmit to long distances the full current required for the lamps, or where it is not advisable to have a large difference of potential between the lamp terminals and earth.

The vector diagrams, figs. 44, 45, and 46, apply as well to the case of a series transformer as to that of a shunt or parallel transformer; only, instead of the pressure E being constant, it is now the current vector *oc* which must be considered to remain of constant length. It will then be seen that variations in the resistance of the secondary circuit lead to no sensible variation in the current; but the secondary E.M.F.— and therefore the primary E.M.F. likewise—will follow the changes in the resistance of the external circuit.

If a transformer is used on a constant pressure circuit, the secondary winding must never be short circuited, or the excessive currents in the coils will burn up the transformer, the assumption being that the latter is not protected by fuses.

A transformer on a constant current circuit must, on the contrary, never have its secondary open-circuited, otherwise the induction in the core will practically reach the saturation limit, and the transformer will be destroyed owing to excessive heating of the iron, even if the insulation does not first break down on account of the abnormally high pressures in both windings. That the induction will increase enormously is evident, since if by breaking the secondary circuit no current is allowed to flow through it, the whole of the primary current will be available for magnetising the core, and the transformer will thus become a very powerful choking coil.

Again, a transformer of which the primary is connected to constant pressure mains takes the least amount of power when the secondary is open-circuited; whereas the power supplied to a *series* transformer is a minimum when its secondary is short-circuited ; for the power absorbed is then only equal to the c^2r losses in the two windings, in addition to the practically negligible amount lost in the core. That the latter quantity is negligible will be evident once it is clearly realised how exceedingly low the induction will be when it is only required to generate sufficient E.M.F. to overcome the resistance of the secondary coil.

44. Magnetic Leakage in Transformers.— In nearly every design of transformer there is a certain amount of magnetism generated by the current in the primary coil which does not thread its way through all the turns of the secondary coil. The amount of this 'leakage' magnetism will evidently increase as the secondary circuit is loaded, the result being that

the ratio of the induced E.M.F.s in the two windings
will not be the same at full load as on open circuit.

Thus the drop in volts at secondary terminals at
full load is generally greater than can be accounted for
by the ohmic resistance of the coils; and as this
increased drop is in many cases objectionable, and the
whole question of magnetic leakage appears to be but
improperly understood, or at least treated with a great
deal of unnecessary complication, it may be advisable
to dwell upon it at some length.

Transformers for working in parallel off constant
pressure mains must be designed for a certain maximum
drop of volts at secondary terminals between no load
and full load; this drop in pressure can rarely be
allowed to exceed $2\frac{1}{2}$ per cent. of the normal secondary
voltage. It follows that since this drop is due partly
to the resistance of the windings and partly to leakage
of the magnetic lines, it is very much more economical
to keep it within the specified limits by reducing the
leakage magnetism to as small an amount as possible,
than by increasing the weight of copper—and therefore
the size of the transformer—in order to keep down the
resistance of the coils.

In fig. 47 is shown a design of transformer which
would have a large amount of leakage. The current
in the secondary coil, s, produces at some particular
moment a magnetising force (as indicated by the
arrow) exactly equal and opposite to the magnetising

force of the main component of the current in the primary coil. The result is that, although this component of the primary current can produce no magnetic flux through the secondary coil, it will generate a considerable amount of magnetism which will leak across the air space between the two coils.

Although the magnetising component of the primary current will also produce a certain amount of leakage magnetism, the amount of this will generally be exceedingly small, and in any case, since this amount will

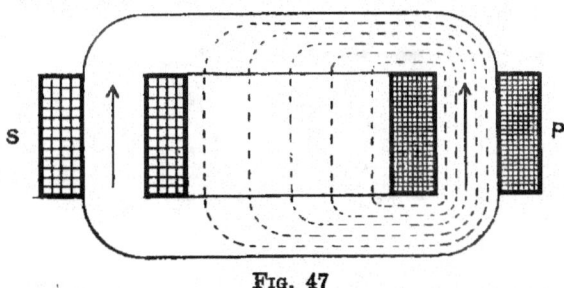

Fɪɢ. 47

be sensibly the same at full load as on open circuit, it will have no appreciable effect in increasing the secondary voltage drop. It should, indeed, be clearly understood that, however great the leakage may be on open circuit, it is only the *increase* of the leakage magnetism *due to the current in the secondary* which has to be considered when studying the question of secondary drop.

Again, with regard to butt joints in the magnetic circuit; although these will always lead to a considerable

increase in the magnetising current, it by no means
follows that they will cause a corresponding increase—
or even the smallest increase—in the amount of the
leakage drop.   It entirely depends upon the position of
the air gaps relatively to the position of the coils.

Returning to fig. 47, it is evident that between
such an arrangement as is there shown, and the case
of an iron ring uniformly wound with two layers of
wire, or a transformer having a *mixed* winding—that is

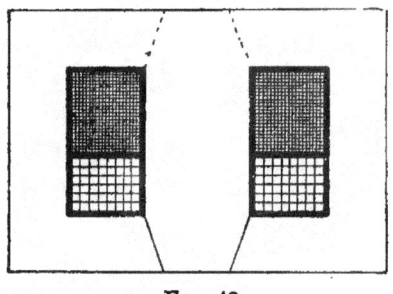

FIG. 48

to say, a winding in which both primary and secondary
coils are wound on at the same time in the same space
—there are many arrangements of coils which will be
more or less successful in getting over this difficulty of
magnetic leakage.

The sections shown in figs. 48 and 49 are respectively those of a Westinghouse and a Mordey
transformer, as tested by Dr. J. A. Fleming in 1892,
from which it will be seen at a glance that the drop
due to magnetic leakage must be considerably greater

in the former than in the latter; and, indeed, the results of Dr. Fleming's tests give a drop of over 1 per cent. for the Westinghouse, while for the Mordey transformer it is only about ·04 per cent.

It will also be readily understood that if such a transformer as the one shown in fig. 49 be made with an 'earth shield,' or metal sheet between the primary and secondary coils, the space taken up by such an arrangement, together with that required for the

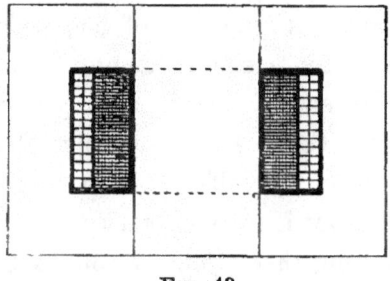

Fig. 49

increased thickness of insulation, will lead to a considerable increase in the amount of the leakage drop. The arrangement of coils adopted by Mr. Gisbert Kapp in his earlier designs of transformers, in which sections of the secondary winding were sandwiched between sections of the primary winding, was an attempt to get something magnetically as good as the quite impracticable 'mixed' winding, and the result was entirely satisfactory; but the amount of space taken up by insulation was, of course, very considerable.

In explaining the effects of magnetic leakage in transformers the assumption is very frequently made that the secondary winding has self-induction. The result is that the vector diagrams of transformers having leakage become somewhat complicated, and, unless this assumption be justified, they must also be, to a certain extent, inaccurate.

Until it can be proved that the secondary winding of an ordinary transformer has appreciable self-induction, we shall, on the contrary, assume that it has none. This will enable the whole question of magnetic leakage to be treated in a more simple manner, and the vector diagrams will be such as can be readily understood.

In order that there may be no misunderstanding as to what is meant by self-induction of the secondary winding, the reader is referred back to fig. 47 (p. 133).

Now, assuming the secondary coil to have no self-induction, the potential difference at its terminals will be equal to the E.M.F. generated in it *due to the magnetism which the primary is able to send through it, minus* the volts lost in overcoming the resistance of the winding. The leakage lines shown in the figure, although they are indirectly due to the current in the secondary coil, do not any of them pass up through the core of the latter, but are all generated by the primary coil.

If, on the other hand, we assume the secondary winding to have self-induction, we must imagine a certain amount of this leakage magnetism—let us say

half of it—to be *generated* by the current in the secondary
coil, and we must therefore conceive of two *independent*
streams of alternating magnetism (with a phase difference of $\frac{1}{4}$ period) passing through this coil.   In other
words, the magnetic flux due to the primary coil
generates a certain E.M.F. in the secondary coil; the
latter also produces a back E.M.F. of self-induction
which, together with the above E.M.F. of *mutual* induction, gives us the *resultant* E.M.F. in the secondary
winding to which the flow of current is due.

When we consider two electro-motive forces, one
of which is somewhat greater than the other, acting
against each other in an electric-circuit, we do not think
of two independent currents flowing in opposite directions through the circuit; but of a single current, the
strength and direction of which depend upon the
resultant or *effective* E.M.F.

Similarly, although it does not necessarily lead to
inaccurate results if we conceive of several independent streams of magnetism in a magnetic circuit, it is
generally preferable, and more correct, to think of a
single resultant flux of induction due to the joint
action of the various magnetising forces at work in the
circuit.

45. **Vector Diagrams of Transformer
having Leakage.**—In order that the diagrams may
not be unnecessarily complicated in appearance, we
shall not at present take into account the voltage drop

which is due to the resistance of the coils; in other words, we shall suppose the ohmic resistance of the coils to be negligible ; the manner in which this may be taken into account has already been explained in connection with fig. 46.

(a). *Secondary Load Non-inductive.*—In fig. 50 let oe_2 represent the secondary E.M.F. on open circuit, due to

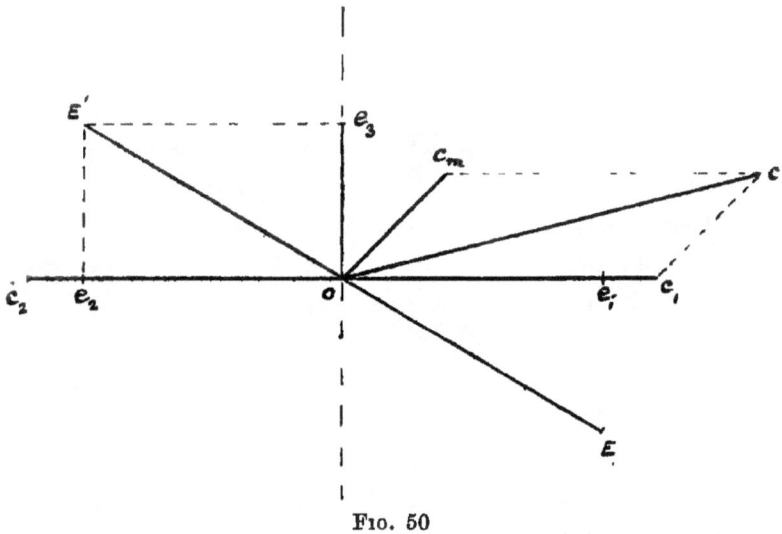

Fɪo. 50

the magnetising current oc_m, the primary E.M.F. being e_1, exactly equal and opposite to e_2.[1]

---

[1] On account of the leakage of magnetism due to the magnetising current oc_m, the primary E.M.F., e_1, would not be exactly equal and opposite to e_2, but would be slightly greater, and at the same time somewhat more than 180° in advance of the secondary E.M.F. The difference between the true position of oe_1 and the position shown in

Let us now suppose that the current c_2 is taken out of the secondary (it will be in phase with e_2, the load being non-inductive and without appreciable capacity), and see in what manner the primary voltage must be varied in order that e_2 may remain as before.

The primary current oc is obtained in the usual way by compounding oc_m and oc_1, the latter component being exactly equal to oc_2.[1] It is this current component, oc_1, which generates the leakage magnetism, and since the latter passes only through the primary coil and not through the secondary, it follows that there will be a component oe_3 of the total primary back E.M.F., which will be $\frac{1}{4}$ period behind oc_1. By compounding e_2 and e_3 we obtain oE', which is the total E.M.F. of self-induction in the primary coil. The applied E.M.F. at primary terminals will therefore be oE, which—since we are neglecting the ohmic drop in the coils themselves—must be drawn exactly equal and opposite to oE'.

The difference between the length of the line oE and the length oe_1 gives us the additional primary volts

the diagram would, however, be so small—even in the case of a transformer having considerable leakage at full load—as to be quite negligible ; and in any case, since this and the following diagrams are drawn for the purpose of studying the *difference* between the primary E.M.F.s at full load and no load, no error would be introduced by assuming e_1 to be equal and opposite to e_2, even if the leakage on open circuit were considerable.

[1] The primary and secondary windings are still supposed to have the same number of turns.

required to compensate for the leakage drop ; or, if we require the percentage leakage drop, this is given by the expression $\dfrac{100(\text{E}-e_1)}{\text{E}}$.

(b) *Self-induction in Secondary Circuit.*—Let us now consider what occurs when the load of the transformer

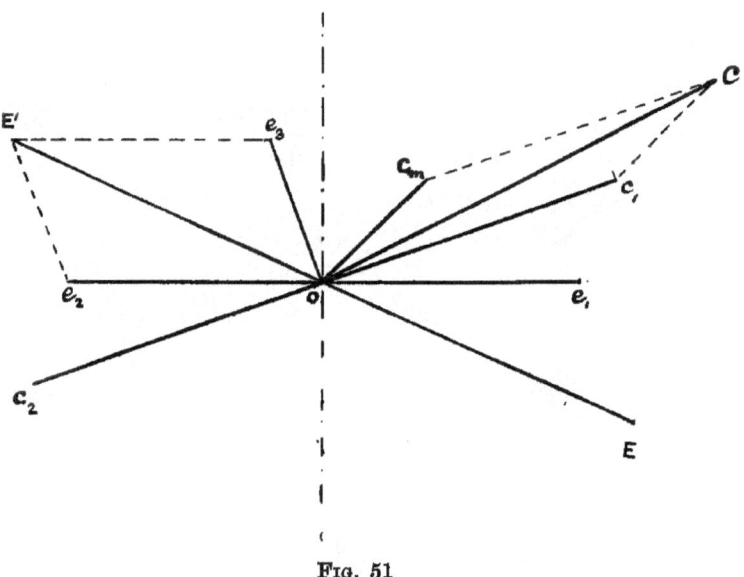

Fig. 51

to which the diagram fig. 50 applies, instead of consisting of glow lamps or a water resistance, is partly inductive. On open circuit the secondary and primary volts (see fig. 51) will be e_2 and e_1 as before; but the vector of the secondary current oc_2, which we will suppose to be of the same length as in fig. 50, is no longer in phase with e_2, but lags behind by an amount depending upon the inductance of the secondary circuit.

The construction for obtaining the primary volts E when the transformer is loaded will be exactly the same as in fig. 50. The main component of the primary current will be, as usual, exactly equal and opposite to oc_2, and the back E.M.F., e_3, due to the leakage magnetism, will be the same as before; its vector oe_3 being drawn at right angles to the current component oc_1. The impressed E.M.F. E will be such as to exactly balance the two components e_2 and e_3 of the total back E.M.F. in the primary coil.

An inspection of the diagram will make it clear why the drop due to leakage in a transformer is *greater* when there is self-induction in the secondary circuit than when the latter is practically non-inductive, as, for instance, when the load consists only of incandescent lamps.

(c) *Capacity in Secondary Circuit.*—The diagram fig. 52, has been drawn on the assumption that the secondary circuit has a certain amount of capacity in addition to its ohmic resistance. The secondary current c_2 will now be in *advance* of the secondary volts e_2, and by constructing the diagram in the same manner as for the last two cases considered we obtain OE as the vector of the impressed primary E.M.F. at full load.

Here, it will be noticed, the amount by which the current c_2 is in advance of the secondary E.M.F. has

been so chosen that the necessary impressed E.M.F. at
full load is actually *less* than that which is required on
open circuit ; and thus, instead of getting a *drop* of
volts at secondary terminals when the primary is on
constant pressure mains, we should expect to get a *rise*
of pressure as current is taken out of the secondary.

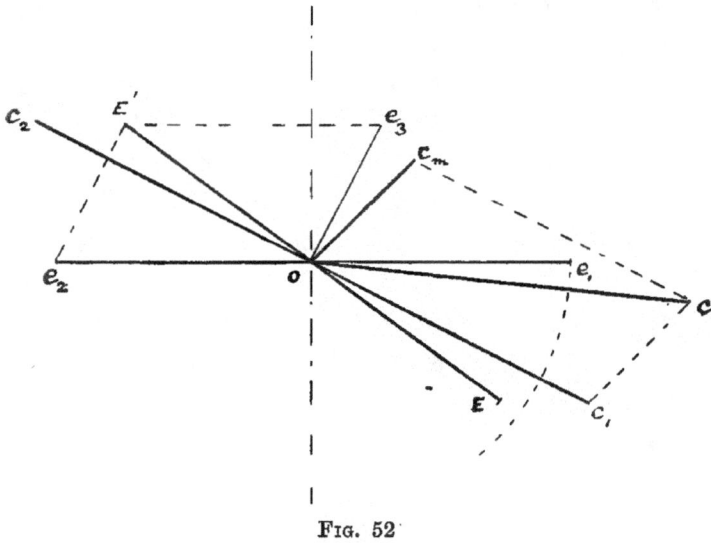

Fig. 52

It is true that we have not taken into account the
voltage drop due to the resistance of the coils ; but it is
evidently only a matter of the relation between the drop
due to resistance and the rise due to magnetic leakage,
which will determine the ratio between the secondary
volts at full load and on open circuit.

As a matter of fact, an actual *rise* in volts at the

secondary terminals of a transformer, the primary of which was on constant pressure mains, has frequently been observed when there has been a certain amount of capacity in the secondary circuit; and this not only in the case of experimental apparatus, but in actual practice, and in connection with commercial transformers, in which neither the leakage nor the resistance of the coils was in any way abnormal.

This phenomenon is often alluded to as one of the many mysterious effects for which electrostatic capacity is held responsible; and when we are told for the first time that capacity in the secondary circuit has the effect of altering the transforming ratio of a transformer, and changing what would otherwise have been a drop into a rise of volts, there is an undoubted charm of novelty about the idea.   But it is wrong to suppose that the effect is due entirely to capacity, as it will only occur if the transformer has an appreciable amount of magnetic leakage; if there is no leakage, neither capacity nor self-induction in the external circuit will prevent the ratio of the induced E.M.F.s in the primary and secondary coils being identical with the ratio of the number of turns in the two windings.

46. **Experimental Determination of Leakage Drop.**—Instead of measuring the voltage at the terminals of a transformer, both on open circuit and at full load—which is often inconvenient—it is evident that we have merely to measure the amount of the back

E.M.F. due to leakage in order that we may determine
the leakage drop at full load.   That is to say, if we can
experimentally determine the length of the vector oe_3
in figs. 50, 51, and 52, we can, by constructing the
complete diagram, predict the amount of the drop at
full load.

By short-circuiting the secondary winding through
an ammeter, the secondary voltage e_2 becomes equal to
zero, and the E.M.F. required at primary terminals
when the current c_2 passes through the secondary coil
will be equal and opposite to e_3.   In making the
experiment it is, of course, important that the frequency should be the same as that of the circuit to
which the transformer is to be connected.

If the drop due to the resistance of the windings is
comparatively large, and it is required to take this into
account, the diagram for the short-circuited transformer
would be as shown in fig. 53.

Here e_2 is the E.M.F. which is required to overcome
the resistance, r, of the secondary winding, together with
that of the ammeter and connections ; it will be in phase
with c_2 and equal to $c_2 \times r$, the assumption being that
the self-induction of the ammeter and connections is
quite negligible.   The induced E.M.F. due to the
leakage magnetism will be e_3 as before ; the total back
E.M.F. in the primary coil being oe'.

The primary current will be oc, which has been
drawn equal and opposite to oc_2 because the magnetising

component of the total current will now be so small that it need not be taken into account. The volts lost owing to resistance of the primary winding will be oE_B, in phase with oc and equal to $c \times R$, where R stands for the resistance of the primary coil. Hence the voltage at primary terminals will be oE, which is obtained by compounding oE_R and a force exactly equal and opposite to oe'.

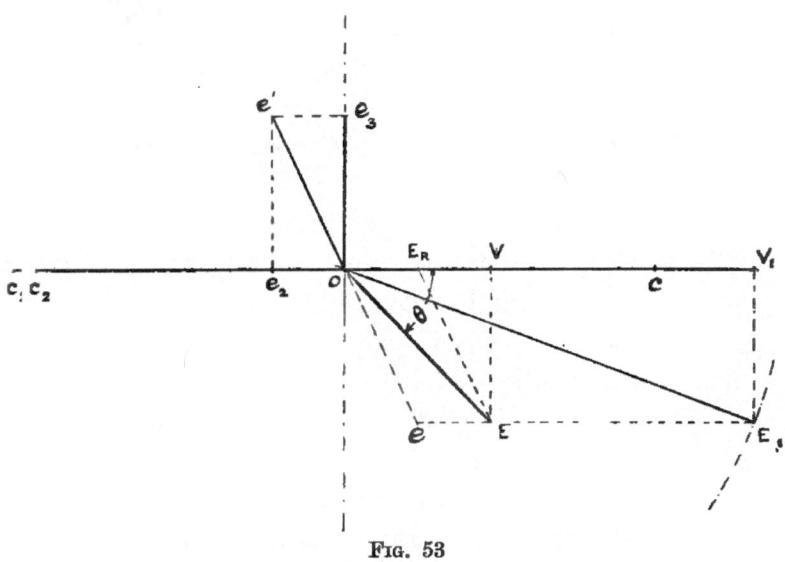

Fig. 53

It follows that, since the 'leakage' component of the total E.M.F. is at right angles to the 'ohmic' component, if $E =$ the necessary E.M.F. required to produce the current c_2 in the short circuited secondary, and $V =$ the E.M.F. required to overcome the resis-

L

tance of the secondary circuit and primary winding, the back E.M.F. (e_3) due to the leakage will be equal to $\sqrt{E^2 - V^2}$.

As an example of the practical use which may be made of this method for determining the drop in pressure at secondary terminals between no load and full load—or, indeed, any intermediate load—let us suppose that it is to be used for determining the percentage drop in a transformer which would absorb an inconveniently large amount of power if tested in the ordinary way by passing the full secondary current through a non-inductive resistance of suitable size.

Instead of putting the ammeter in the secondary circuit, it will be generally more convenient to put it in the primary circuit, the secondary terminals being merely short-circuited by means of a heavy copper connecting piece. In addition to the ammeter in the primary circuit, there should also be a wattmeter to measure the actual power supplied to the transformer, and a voltmeter to indicate the potential difference at the terminals.

The transformer being connected to a source of supply of the correct frequency, the voltage at the primary terminals is so adjusted that the required current is indicated by the ammeter. The corresponding readings of the wattmeter and voltmeter are now taken.

Let us suppose the three readings to be:

Current through primary coil = 10 amperes.

Potential difference at terminals = 30 volts.

Power supplied = 180 watts.

The reading of the wattmeter (180 watts) gives us at once the amount of the copper losses corresponding to a particular value (10 amperes) of the primary current; because, as we have already seen, the induction in the core is so low that we are quite justified in assuming the watts lost in magnetising the iron to be negligible. If, therefore, we divide 180 by 10 we obtain the number 18 as the amount of that component of the total applied E.M.F. which is in phase with the current (see the vector ov in fig. 53).

The 'power factor' being equal to $\dfrac{180}{30 \times 10} = {\cdot}6$, which is very nearly equal to the cosine of 53°, it follows that the vector of the total applied E.M.F. ($oE = 30$ volts, fig. 53) must be drawn at an angle of 53° in advance of oc, which represents the primary current. The 'leakage component' (oe_3) of the total E.M.F. is evidently equal to $EV = \sqrt{30^2 - 18^2} = 24$.

Having ascertained the value of e_3, the drop at secondary terminals due to magnetic leakage can now be determined in the manner explained in § 45 (see fig. 50, p. 138); or, by carrying the construction of the diagram fig. 53 a step further, we may determine the total percentage drop in the following way.

Produce the dotted line eE (parallel to oc) in the direction E_p and, from o as a centre, and with a radius oE_p equal to the normal amount of the primary impressed E.M.F., describe an arc which cuts the above line at E_p. Join oE_p, and project on to oc. Then, as a careful examination of the diagram will show, the percentage drop at secondary terminals when the secondary current is equal to $oc \times$ ratio of turns in the two coils, and the transformer is working on non-inductive load, will be given by the expression:

$$\frac{100 \times (oE_p - vv_1)}{oE_p}$$

47. Determination of Leakage Drop on Open Circuit.—Although in the majority of transformers the leakage due to the magnetising current is exceedingly small, there are certain cases in which the ratio of primary and secondary E.M.F.s on open circuit is by no means identical with the ratio of turns, and it is often useful to be able to determine experimentally the amount of the voltage drop due to the leakage on open circuit.

Let q stand for the ratio of the effective number of turns in the two windings; then, if there is no magnetic leakage on open circuit:

$$e_p = q \times e_s$$

where e_p and e_s stand for the induced E.M.F.s in the primary and secondary coils respectively. Thus if we supply e_s volts to the secondary terminals, the volts indicated by a voltmeter connected across the primary terminals would be equal to $e_s \times q$, because the very small ohmic drop due to the magnetising current need not be taken into account. If, however, there is leakage, the volts at primary terminals will not be $e_s \times q$, but $e_s \times q \times m$, where m is a multiplier depending upon the amount of the leakage and which will always be less than unity.

Similarly, if we now supply e'_p volts to the primary terminals, the voltage at secondary terminals will be $e'_p \times \dfrac{1}{q} \times m$.

By eliminating m from the two equations,

$$e_p = e_s \times q \times m$$

$$\text{and } e'_s = e'_p \times \frac{1}{q} \times m,$$

we obtain the expression :

$$q^2 = \frac{e_p e'_p}{e_s e'_s} \qquad . \qquad . \qquad . \qquad (30).$$

In this manner the ratio of the number of turns in the two windings can be experimentally determined ; and, knowing this, the open circuit drop in volts can of course easily be ascertained.

As an example let us suppose that, by transforming up from 102 volts on secondary, we obtain exactly 2,000 volts at primary terminals, and that by transforming down from 2,000 volts we get only 98 volts at secondary terminals, then

$$q^2 = \frac{(2,000)^2}{102 \times 98}$$

from which we get

$$q = \frac{2,000}{99\cdot8} = 20\cdot04,$$

and the percentage drop on open circuit is equal to

$$\frac{99\cdot8 - 98}{99\cdot8} = 1\cdot8 \text{ per cent.}$$

It is hardly necessary to point out that, unless the open circuit drop is considerable, the ratio of turns may be written $q = \dfrac{e_p + e'_p}{e_s + e'_s}$, because this is very nearly equal to the square root of $\dfrac{e_p e'_p}{e_s e'_s}$ when the difference between e_p and e'_p and between e_s and e'_s is very small.

If we transform up to the normal voltage of the supply circuit and then transform down again from the same voltage—as was done in the numerical example which has just been worked out—the ratio q may be written $\dfrac{2e_p}{e_s + e'_s}$, always assuming the difference between e_s and e'_s to be small.

48. **Transformers with Large Leakage, for Arc Lamp Circuits.**—When arc lamps are run in parallel off constant-pressure mains, it is usual to insert a choking coil in series with each lamp, or group of lamps, in order to keep the current as nearly constant as possible. Another way of preventing considerable variations in the current consists in connecting the arc lamps directly to the terminals of a transformer which

has a large amount of magnetic leakage.   The diagram
for such an arrangement is shown in fig. 54, which is
almost exactly the same as fig. 50 (p. 138), only the back
E.M.F. due to leakage has been proportionately increased.

So as not to complicate the diagram, it has been
assumed that there is no self-induction in the arc lamp

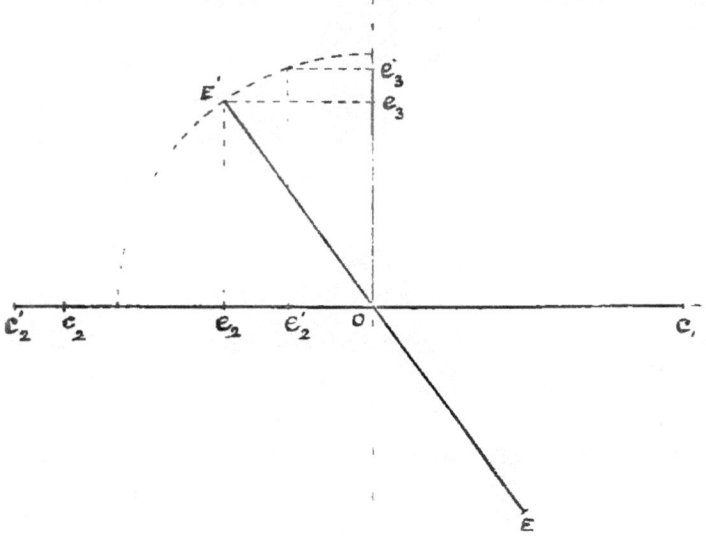

Fig. 54

circuit.   The volts e_2 at secondary terminals will therefore be in phase with the secondary current c_2.   The
back E.M.F. due to leakage is e_3, which, it must be
remembered, is proportional to the current c_2, and the
total induced E.M.F. in the primary coil is E'.

Since the primary pressure is constant, the point E'

will lie on the dotted circle described from the centre o, whatever may be the resistance of the external circuit. Let us suppose this to be halved; the secondary current will instantly increase to c'_2, thus bringing the leakage volts up to e'_3 and reducing the effective volts to e'_2; the relation between c'_2 and c_2 being such that e'_2/c'_2 is only half as great as e_2/c_2.

A glance at the diagram will show that, owing to the large amount of the magnetic leakage, only a very small increase in the current is required in order to considerably reduce the effective E.M.F. in the secondary circuit.

49. Maximum Output of Transformer having Large Magnetic Leakage.—Referring again to fig. 54, it will be seen that the output of the transformer, $e_2 \times c_2$, may be written $e_2 \times ke_3$, where k is a constant, because the leakage volts e_3 will be almost exactly proportional to the current c_2. It follows that the transformer will be working at its greatest output when the product $e_2 \times e_3$ is a maximum; and since the primary E.M.F. E (and therefore also the back E.M.F. E') is supposed to remain constant, this maximum will occur when the rectangle $oe_2E'e_3$ is a square, or when

$$e_2 = e_3 = \frac{E}{\sqrt{2}},$$ the assumption being that the resistance of

the primary coil is negligible.

50. General Conclusions regarding Magnetic Leakage in Transformers.—Let $B_s =$ the maximum value of that component of the total induction in primary core which passes also through the secondary coil, and B_l = the maximum value of the leakage component of the total induction in primary core.

Then, assuming the resistance of the path of the leakage lines to be constant, we may write :

B_l ∝ ampere-turns tending to produce leakage,

$$\propto T \times c_1$$

where T = the number of turns in the primary coil, and c_1 = the main component of the primary current. And since the mean back E.M.F. in primary, due to leakage (the vector oe_3 in the last few diagrams), is proportional to $B_l \times T \times n$, where n = the frequency, it follows that we may write

$$e_3 \propto T^2 c_1 n \qquad . \qquad . \qquad . \qquad (31),$$

from which we see that the mean value of the ' leakage volts' for any given arrangement of the magnetic circuit is proportional to the square of the number of turns in the winding, and to the current, and to the frequency.

But as it is not the actual value of the leakage E.M.F. with which we are generally concerned, it will be advisable to express this as a percentage of the mean E.M.F. available for generating current in the secondary circuit. This last will be proportional to $B_s \times T \times n$; and if we divide the actual ' leakage E.M.F.' by this amount, we obtain the expression :

Percentage back E.M.F. due to leakage

$$\propto \frac{T^2 c_1 n}{\mathbf{B}_s T n} \propto \frac{T c_1}{\mathbf{B}_s}. \qquad . \qquad (32)$$

It follows from this that if we consider an ordinary transformer on a non-inductive load, the *percentage* drop in volts due to leakage is independent of the frequency, but is inversely proportional to the induction through the secondary coil. The improvement from the point of view of leakage drop, although it is perhaps not generally realised, is very noticeable as higher inductions are used in the cores of transformers; and if a transformer, designed to work on a constant-pressure circuit at a frequency of 100, is connected to a circuit of which the frequency is 50, there is a decided reduction in the percentage amount of the leakage drop—not, however, because of the reduction of frequency *per se*, but because the magnetic flux through the secondary coil has increased, whereas the leakage magnetism has remained as before.

In § 45 (p. 137) it was shown how the amount of the leakage drop depends upon the nature of the external circuit; and in fig. 51, where the load is assumed to be partly inductive, the drop was shown to be greater than in fig. 50, where the load is non-inductive. In fig. 55 two curves have been plotted from measurements made on an actual transformer having considerable leakage.

The upper (full line) curve shows the relation between

secondary volts and current when the transformer is on a non-inductive load—the primary volts being kept constant—whereas the lower (dotted) curve was

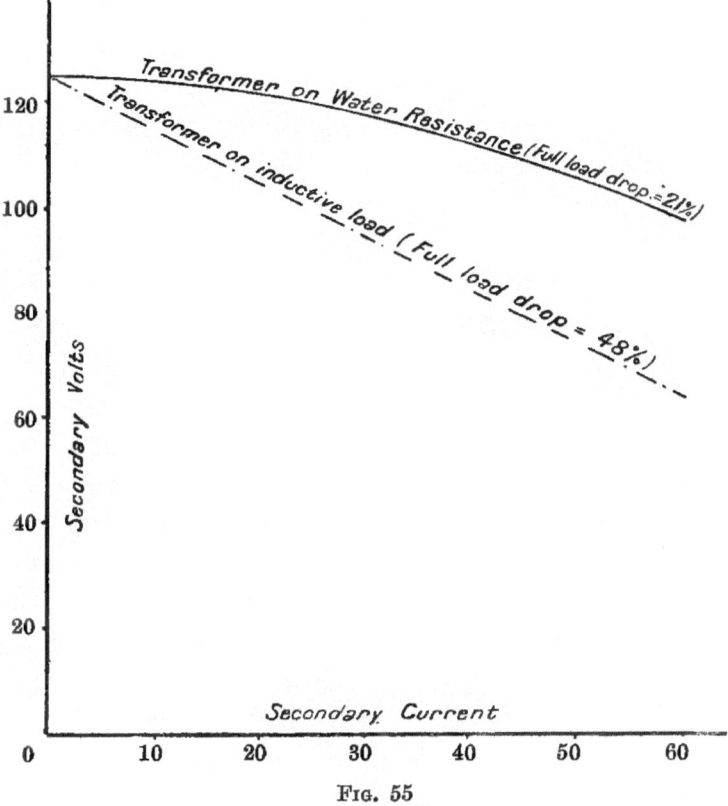

Fig. 55

obtained with an external circuit of very large self-induction and almost negligible resistance. In this case the current vector oc_2 lags nearly 90° behind oe_2 (see figs. 50 and 51), thus bringing e_3 almost exactly in phase with e_2. The result is that the leakage E.M.F. being now opposed to the primary impressed E.M.F., the voltage drop at secondary terminals will be directly proportional to the current, which accounts for the dotted curve in fig. 55 being a straight line.

In a paper read by Dr. G. Roessler, in July 1895, it was pointed out that the leakage drop in a transformer is greater when the curve of the applied E.M.F. is of a peaked form than when it is more rounded or flatter. This is explained by the fact that the 'wave constant' (see p. 37) is greater for the peaked form of wave than for the flat form, which means that, for a given $\sqrt{\text{mean}}$ square value of the E.M.F., the maximum induction—which is proportional to the area or *mean* value of the wave—will be lower in the former case than in the latter; and this, as we have already seen in connection with equation (32), will lead to a slight increase in the amount of the leakage drop.

As the *efficiency of transformers* is a matter of great importance in practice, we shall now turn our attention to the causes of loss of power, both in the laminated iron cores and in the copper conductors of which the windings are composed.

51. Losses Occurring in the Iron Core.— In §§ 23 to 26 the losses due both to eddy currents and to hysteresis which occur in the iron cores of choking coils or transformers were discussed at some length;

but there are certain conditions which influence the amount of these losses, and it is proposed to deal briefly with these in the present paragraph.

(a) *Effect of Wave Form on the Hysteresis Losses.*—In the admirable paper read by Messrs. Beeton, Taylor, and Barr on May 14, 1896, before the Institution of Electrical Engineers, the influence of the shape of the applied potential difference wave on the iron losses of transformers was very clearly explained, and the results of a large number of tests made by the authors served to prove (1) that the eddy current losses are independent of the wave form of the applied E.M.F., provided the $\sqrt{\text{mean square}}$ value of this E.M.F. remains constant; and (2) that the hysteresis losses for a given $\sqrt{\text{mean}}$ square value of the applied E.M.F. are dependent only upon the *mean* value of the E.M.F., or on the area of the E.M.F. wave; and variations in the shape of the wave will only produce alterations in the hysteresis losses if the ratio $\dfrac{\sqrt{\text{mean square value of E.M.F.}}}{\text{mean value of E.M.F.}}$ is altered.

The above ratio is what we have called the 'wave constant' in § 15, p. 37 (to which the reader is referred), and as the induction in the iron is proportional to the *mean* E.M.F., or to the *area* of the E.M.F. wave, it follows that the hysteresis losses, for a given frequency and E.M.F., will depend upon the value of this 'wave constant'; and since, as was pointed out at the end of

§ 50, the latter quantity is greater for a peaked form of wave than for a flat form, the hysteresis losses in a transformer will be somewhat less when the supply is taken from an alternator giving a pointed or peaked wave than when it is taken from a machine giving a more rectangular-shaped wave.

The greatest difference in the total iron losses due to variations in the shapes of waves which Messrs. Beeton, Taylor, and Barr obtained in their experiments amounts to about 18 per cent. Tests made by Mr. Steinmetz as long ago as 1891 showed that the core losses in a transformer when worked with the distorted wave of an 'iron clad' alternator were 9 per cent. less than when worked with the sine wave produced by an alternator having no iron in its armature.

With regard to the fact that the eddy current losses, as proved by Messrs. Beeton, Taylor, and Barr, are independent of either the wave constant or the actual shape of the E.M.F. curve, this is only what one would expect, because, as was explained in § 23 (p. 57), these losses are proportional to the square of the induced E.M.F., and have therefore nothing to do with the *mean* value of the E.M.F., or with the actual maximum value of the induction.

(*b*) *Effect of Temperature on the Iron Losses.*—If the open circuit losses in a transformer are measured both when the transformer is cold and again after it has been warmed up, through having been left on the supply for a few hours, it will be found that with the rise in temperature of the iron core the losses in the latter have been somewhat reduced. This is due to the fact that the electrical resistance of the iron has increased, thus causing a reduction in the eddy current losses. The hysteresis losses will remain practically constant, and the saving in the total losses can only be attributed to the reduction of the eddy currents.

In a transformer tested by Mr. E. Wilson the reduction in the open circuit losses when the temperature was raised from 16° to 50° C. was 4 per cent. Another transformer, tested by Dr. J. A. Fleming, showed a difference of nearly 10 per cent. in the core loss between the amount obtained from measurements made when the transformer was cold and again when it had reached its final temperature, after being connected to the source of supply for a sufficient number of hours.

There is another, and entirely distinct effect of temperature on the core losses of transformers. It is found that in many cases the iron losses increase very considerably with time; that is to say, if the watts lost on open circuit are measured immediately after completion, and again after the transformer has been connected to the circuit for a period of two or three months, it will be found that, in certain cases, the losses have increased as much as 50 or even 100 per cent.

This 'ageing' of the iron, which leads to such considerable variations in the hysteresis losses, is the result of the temperature rise due to the dissipation of energy in the transformer; but it does not occur with every sample of iron. Sheet iron for transformers can now be obtained which shows no appreciable increase in the hysteresis losses, even when kept at comparatively high temperatures for a considerable time. No satisfactory explanation of this phenomenon has yet been given.

(c) *Reduction of Core Losses when Current is taken from Transformer.*—It has long been known that when current is taken from an alternator having a laminated iron core, the total loss of power in the core is very often considerably less than when the machine is running on open circuit. That the same thing occurs in the case of transformers, when the latter have an appreciable amount of magnetic leakage, is more than probable; and some experiments made by Mr. E. Wilson (see 'The Electrician,' February 15, 1895) on a two-kilowatt Westinghouse transformer—in which, as was previously pointed out (see p. 134), the coils are so arranged as to give rise to a comparatively large amount of magnetic leakage—would seem to confirm this.

However, since no satisfactory explanation of this effect has yet been given, any further discussion of the question in these pages would be out of place. It will be sufficient if the reader will refer to the curves of fig. 56, which are plotted from the results of some preliminary tests made by Mr. Edward W. Cowan and the author in 1895 on the core losses of a small experimental alternator. Here the reduction of the power lost other than the c^2r loss in the coils is seen to diminish very rapidly as current is taken from the armature,

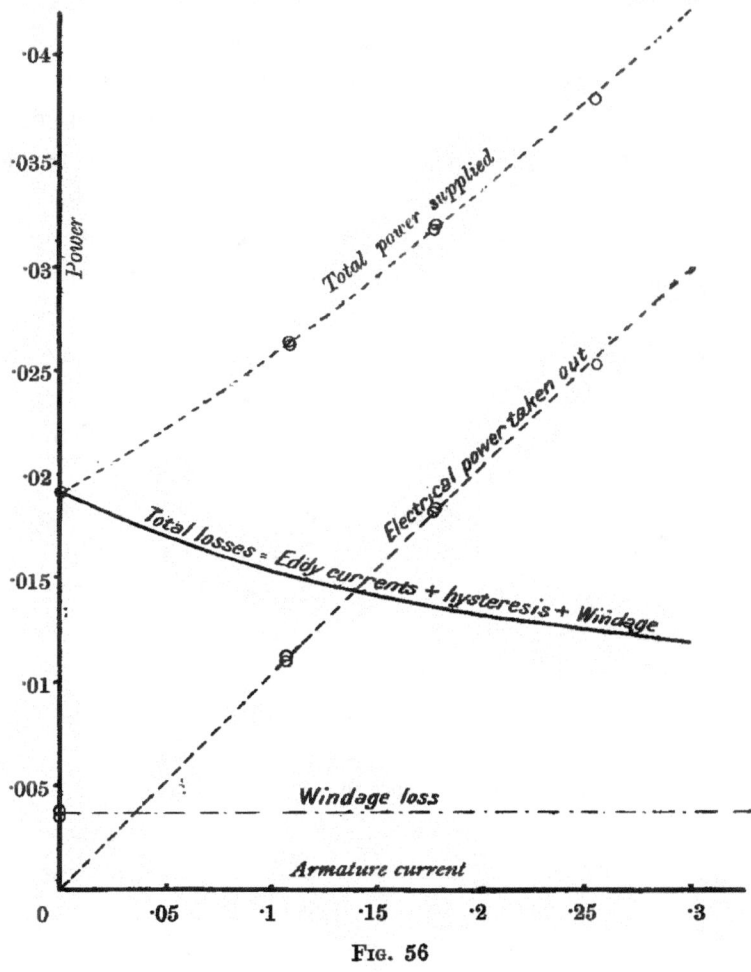

Fig. 56

M

Want of time, owing to pressure of other work, prevented the investigation being carried any further.

52. Losses in the Copper Windings.—Apart from the c^2r losses in the coils, which may be easily calculated for any load provided the resistances of the two windings are known, there are one or two causes which tend to increase the amount of these calculated losses, and, although they are of relatively small importance, they should not be entirely overlooked. And in any case it must not be forgotten that the copper losses for a given output may be appreciably greater after the transformer has been at work for some hours, and the coils have had time to warm through, than if measured just after the load has been switched on, and before the temperature of the copper has increased to any appreciable extent.

An increase in the resistance of the coils of 1 per cent. for every $2\cdot6^\circ$ C. or $4\frac{3}{4}^\circ$ F. should be allowed for.

(a) *Eddy Currents in the Windings due to Leakage Magnetism.*—Generally speaking, the losses in a closed magnetic circuit transformer due to eddy currents in the coils, which are the direct result of magnetic leakage, are so small as to be quite negligible. It must, however, not be forgotten that in the case of a transformer having considerable magnetic leakage, and wound with solid conductors of large cross-section, there may very well be an appreciable loss of power owing to this cause. But with the laminated or stranded conductors, which it is generally advisable to use for other reasons, a transformer with a closed iron circuit may be compared with an armature of which the conductors are bedded in slots or tunnels, and in which, as is well known, the losses due to eddy currents, even with solid conductors of large cross-section, do not appreciably lower the efficiency of the machine.

(b) *Apparent Increase in Resistance due to uneven Distribution of the Current in heavy Conductors.*—In winding transformers for large currents, it is customary, on account of the difficulties of winding, to use either stranded conductors, or copper tape, of which the thickness rarely exceeds $\frac{1}{8}$-inch. A number of these cables or flat strips may, if necessary, be connected in parallel.

In § 38 (p. 113) the apparent increase in the resistance of a conductor when the current, instead of being continuous and of constant strength, is an alternating one of comparatively high frequency, was briefly discussed; and in the case of transformers—especially if the coils are wound with a large number of layers—the amount of material which may be useless, or even worse than useless, if the conductors are very thick in a direction perpendicular to the surface of the layers might lead to a considerable reduction of the permissible output unless the frequency of the supply is very low.

It is therefore advisable, in winding transformers

M 2

for heavy currents, to use a number of flat strips placed one above the other, and taped together as they are being wound on. It is not necessary that each strip should be thoroughly insulated from its neighbour, as the difference of potential tending to produce uneven distribution of the current is very small, and a thin coating of varnish, or even the slightly oxidised surface of the conductors themselves, will be quite sufficient to prevent any increase in the losses owing to this cause.

As a matter of fact, the sub-division of heavy conductors necessitated on account of the difficulties which

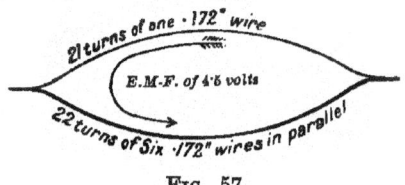

FIG. 57

would be experienced in winding with solid copper of large cross-section renders trouble due to these mutual actions between conductors, or portions of a conductor, unlikely to arise in practice.

53. Effect of Connecting Windings of different Numbers of Turns in Parallel.—One result of winding transformers with several wires or strips in parallel is that mistakes are liable to be made in counting the number of turns in the different layers ; this would lead to there being a certain difference between the E.M.F.s generated in the separate conductors which, when joined together at the terminals, constitute one of the windings ; and the circulation of current due to this cause would not only result in an increase of the current supplied to the transformer, but the power dissipated in the two windings might lead to a very considerable reduction of the efficiency.

Let us, for instance, consider a transformer designed for a ratio of 2,000 to 100, and of which the secondary coil consists of 22 turns of 7 wires in parallel, each wire

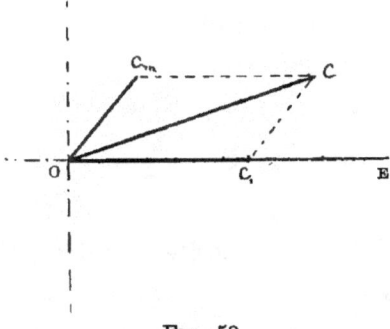

Fɪɢ. 58

being ·172″ in diameter. The number of turns in the primary winding is 436, and the open circuit magnetising current is ·134 amperes, while the power supplied to the transformer is 190 watts. Hence the ‘ power factor ’ is ·71 ; and since this is approximately equal to cos. 45°, it follows that, in fig. 58, if oe is the vector of the primary impressed E.M.F. and oc_m that of the magnetising current ($= ·134$), the angle $c_m$oe will have to be drawn equal to 45°.

Now suppose that one of the seven wires of the secondary is wound with one turn short—i.e. with 21 turns instead of 22. There will be a difference of potential of $100 - \dfrac{100 \times 21}{22} = 4 \cdot 5$ volts, tending to produce a flow of current through the secondary coils.

Assume the average length per turn of the secondary winding to be $5\frac{1}{2}$ feet; then the ohmic resistance of the circuit through which the current will flow is due to $21 \times 5\frac{1}{2} = 115 \cdot 5$ feet of $\cdot 172''$ copper wire, in series with $22 \times 5\frac{1}{2} = 121$ feet of $6 - \cdot 172''$ copper wires in parallel, or about $\cdot 048$ ohms.

The current due to the E.M.F. of $4 \cdot 5$ volts (see fig. 57) will therefore be $\dfrac{4 \cdot 5}{\cdot 048} = 94$ amperes, and as this current flows round the core 22 times in one direction and 21 times in the opposite direction, the resultant ampere-turns tending to demagnetise the core will be 94. This will give rise to a component of the primary current in phase with E, equal to $\dfrac{94}{436} = \cdot 216$ amperes, which is represented in the diagram, fig. 58, by the distance oc_1. By compounding the current vectors c_m and c_1, the resulting primary current oc is obtained. In this case it will be found to be equal to $\cdot 315$ amperes, as against $\cdot 134$ amperes, which is the true magnetising current required by the transformer.

54. Experimental Determination of the Copper Losses.—In connection with the experimental determination of the voltage drop in transformers, it has already been pointed out (see p. 147) that a watt-meter placed in the primary circuit will correctly measure the total amount of the copper losses if the secondary is short-circuited, and the pressure at the primary terminals so adjusted that the required current passes through the coils.

In fig. 59, **T** is the transformer to be tested. Its secondary coil is short-circuited through the ammeter **A**, which must be capable of indicating currents at

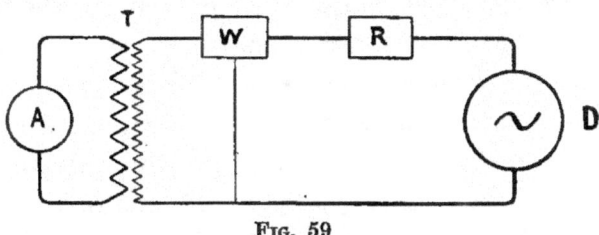

Fig. 59

least equal to the normal secondary current of the transformer at full load. **W** is the wattmeter in primary circuit; **D** the source of supply, of which the frequency must be the same as that of the circuit on which the transformer is intended to work; and **R** is an adjustable resistance, choking coil, or regulating trans-former, which permits of the voltage at primary ter_minals being so varied as to send any desired current through the ammeter **A**.

Since the maximum induction in the core of the transformer will only be of such an amount as to generate an E.M.F. in the secondary coil sufficient to overcome the resistance of the winding, it is evident that the hysteresis and eddy current losses which occur in the iron when the transformer is working at its normal voltage are not included in the reading of the watt-meter, and the latter may therefore be said to measure the copper losses only.

It is true that the wattmeter reading will also include any eddy current losses, either in the stampings or metal castings, due to the leakage magnetic lines ; but as these are losses which occur only when the trans-former is loaded, and which must be reckoned over and above the core losses proper—which are due to the magnetic flux required to generate the normal E.M.F. in the secondary coil—we are justified in including any such additional losses, due to the current taken out of the transformer, in what is generally understood when we speak of the copper loss.

55. Efficiency of Transformers.—If c_2 and e_2 stand respectively for the secondary current, and E.M.F. at secondary terminals, the efficiency of a transformer at any output is given by the expression :

$$\frac{e_2 c_2}{e_2 c_2 + \text{watts lost in iron} + \text{watts lost in copper}}.$$

Assuming the copper losses to increase as the square of the output, and the iron losses to remain constant, it can be mathematically deduced from the above expression that the most economical load for a transformer is that which makes the copper losses equal to the iron losses; and this explains why, in many transformers having a comparatively small iron loss, the maximum efficiency occurs before the full load is reached.

The table on p. 170 is intended to give some idea of the permissible losses in transformers of various sizes, the primary voltage being, say, 2,000, and the frequency about 60 complete periods per second.

The figures are approximate only, and apply to parallel or shunt transformers, such as are used on electric lighting circuits; attempts to make commercial transformers in which the total losses in iron and copper are much below the amounts given in the table will only lead to a large and uneconomical increase in the cost of production.

There can be no doubt that there is still much room for improvement in transformers, and a few years hence —probably owing to further reductions of the hysteresis losses in the iron—we may hope to find their efficiency at light loads considerably improved ; but at the present time it is false economy to reduce the losses in transformers to an ideally small amount; and the many specifications issued without due consideration, by which young and over-zealous engineers attempt to obtain transformers of unreasonably high efficiencies, cannot be said to serve any useful purpose. A transformer should never be considered singly, but always in connection with the whole system of which it forms a part. In dealing with the question of efficiency it is, therefore, necessary to consider, on the one hand the interest and depreciation on the first cost of the transformer ; and, on the other hand, the effect of any reduction of the iron (or copper) loss upon the all-day or all-year coal consumption at the generating station.

TABLE GIVING THE IRON AND COPPER LOSSES WHICH MAY BE ALLOWED IN GOOD TRANSFORMERS OF VARIOUS SIZES

Maximum output in kilowatts	Watts lost in the iron on open circuit	Watts lost in the copper at full load
1	43	22
1·5	50	32
3	70	65
4·5	90	90
6	105	115
9	135	165
12	165	205
15	185	245
20	220	300
25	250	345
30	280	390

56. **Temperature Rise.**—The relation between the cooling surface of a transformer and the total power dissipated in the form of heat should be such that the final temperature attained after the transformer has been working for a sufficient length of time at its rated output shall not be injurious to the insulation.

The 'temperature drop,' due to increase of the resistance of the coils as these are warmed up, must also be taken into account ; and the question of 'ageing '

of the iron in the core is one which must likewise be
carefully considered. The permissible rise in temperature of any given transformer is therefore a question
which must be dealt with by the manufacturer.
Generally speaking, the temperature rise should be as
high as possible in order to reduce first cost.

Besides the increase in the E.M.F. lost in overcoming ohmic resistance, there will be, of course, a
proportional increase of the c^2r losses, which must be
duly taken into account; the saving in the core loss
owing to the greater resistance of the iron and conse-
quent reduction of the eddy currents will not generally
be found to compensate for the larger amount of the
copper loss.

In order to illustrate the effect of temperature on
the core losses, a rough test was made on a transformer
having a large amount of magnetic leakage.

On the assumption that the watts lost in the core
would be considerably reduced, we should expect to find
(1) a decrease in the open circuit magnetising current as
the transformer is warmed up, and (2) an improvement
in the transforming ratio, owing to a slight reduction of
the leakage drop, corresponding to the decrease in the
magnetising current.

A glance at fig. 60 will show that the experiment proved this to be the case. The primary of the
transformer was connected to a source of supply of
which the pressure and frequency were kept as con-

stant as possible ; and, as the temperature rose, readings were taken at intervals of the primary current,

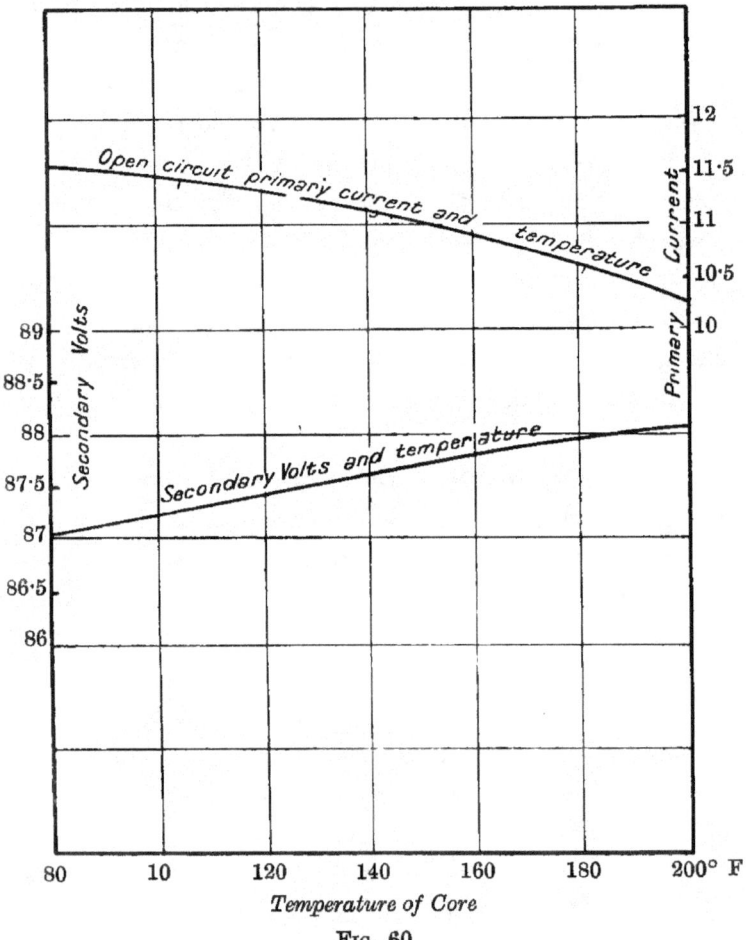

Fig. 60

secondary volts (open circuit), and temperature of the
core.

The two curves in fig. 60, which represent the mean values of a large number of readings, show not only a considerable reduction in the magnetising current at the higher temperatures, but also a quite appreciable increase in the secondary E.M.F. with constant primary potential difference.

The amount of cooling surface which should be allowed per watt lost in the transformer in order that the temperature rise may not be excessive should be experimentally determined, as no reliable rules can be given which are applicable to all designs of transformers.

Then, again, exactly what is meant by the cooling surface is not very easy to define. A transformer which is enclosed in a cast-iron case will naturally reach a higher temperature than a similar transformer which is not enclosed in a case, and to which the surrounding air has free access. If the case is filled with oil, the transformer will keep cooler; but this method of insulation, although it has been used in America, has not proved successful when tried in this country.

As a rough approximation, a cooling surface of from 4 to 5 square inches per watt lost should be allowed in order to keep the temperature rise within reasonable limits.

According to Mr. Gisbert Kapp, who has made a number of tests on the heating of transformers, the final temperature rise in degrees Fahrenheit may be taken as equal to $\dfrac{\text{Watts lost}}{\text{Cooling surface in square inches}} \times 430,$ for

a total rise in temperature over the surrounding air of about 110° F. The transformers tested were all enclosed in watertight iron cases, which were placed on a concrete floor in a large covered space.

From tests made by the author on a three kilowatt transformer also enclosed in iron case, the multiplier in the above expression, instead of being 430 as found by Mr. Kapp, amounted to as much as 600 ; but the space in which this transformer was tested was probably more confined than in the experiments made by Mr. Kapp, in which the air had free access to all parts of the cases in which the transformers were placed.

With regard to what has been called the final temperature of a transformer, it must ·not be supposed that this is by any means equivalent to the maximum temperature attained by the majority of transformers in practice, even when these are fully loaded or even overloaded at certain times of the day. The greatest output of a transformer on an electric lighting circuit is not likely to be required for more than an hour or two each day ; and as the watts lost in the copper are proportional to the square of the output, it follows that the moment there is any reduction in the load, the power wasted in heat begins to diminish at a still more rapid rate. The result is that in actual practice the maximum temperature rise in a transformer is usually very much less than the final temperature rise which would be reached if the transformer were worked continuously at full load. A good transformer should be capable of standing an overload of one and a half times or even twice its rated output for short periods of time, without showing any signs of injurious heating. The voltage drop will, however, in this case generally be excessive.

That a transformer may get hot and yet have a high efficiency is a fact which some people appear to be unable to accept.

The man who considers that a temperature test is the best method of determining the relative merits of different designs of transformers is still to be met with; and this may account for the many absurd temperature clauses which one comes across in transformer specifications. For it would be unquestionably absurd to greatly increase the amount of material in a transformer for the sole purpose of limiting the rise of temperature to, say, 40 or 50° F., when a rise of 80 to 100° would probably still fall short of producing a maximum temperature likely to be injurious to any of the materials of which the transformer is composed.

That there is really no connection between temperature rise and efficiency of a transformer should be evident when it is considered that the temperature depends, not only upon the heat generated, but also upon the cooling surface through which this heat can get away; and indeed it often happens that, of two transformers of the same output, but of different designs, the one in which the greater amount of power is dissipated will have the smaller temperature rise.

The following results of tests made on the iron heating of two 3,000-watt transformers, supplied by different makers, should be of interest. The first transformer, which we will call A, attained a final temperature of 48° F. after being left on the supply circuit for over twelve hours. The second transformer B, which was treated in exactly the same manner, attained a final temperature of 73° F. Careful measurements were then made of the power lost in the cores, and this was found to be for the transformer A 100 watts, whereas for B it was only 71 watts.

57. **Effect of taking a small Current from a Transformer having considerable Primary Resistance.**—Although the following considerations are not likely to be of much practical importance, it is interesting to note that under certain conditions the impedance of the primary winding of a transformer may *increase* when current is first taken out of the secondary winding. In other words, if we suppose the potential difference at primary terminals to remain constant, the primary current, when the secondary is lightly

loaded, may in certain cases be *less* than when the secondary is open-circuited.

In § 37 (p. 109) the effect of shunting a choking coil with a non-inductive resistance was discussed at some length, and although in this case it is proposed to adopt a somewhat different geometrical construction to that which was used in fig. 39, the problem with which we are at present concerned is really very similar to the one referred to.

In the first place, we must suppose the primary winding of the transformer to have a considerably higher resistance than is usually the case in practice; that is to say, the E.M.F. lost in overcoming ohmic resistance, even when the secondary is open-circuited, must be fairly large in comparison with the E.M.F. of self-induction; and, in the second place, we will assume, partly in order to simplify the diagram and partly to increase the effect which we are investigating, that there is no iron in the core of the transformer.

In fig. 61 let oc_m be the open circuit primary current, and oe the impressed primary E.M.F. This last, it will be seen, is made up of oe_{cm} in phase with the current and representing the E.M.F. required to overcome the resistance of the primary coil, and oe, one quarter period in advance of the current and exactly equal and opposite to the E.M.F. of self-induction generated in the primary.

Suppose, now, that a small current c_2 is taken out

N

of the secondary through a non-inductive resistance. There will be a balancing component c_1 in the primary, which will require an E.M.F. e'_{c1} to force it against the resistance of the coil.

In order to obtain the magnetising component c_o of the new primary current, describe the dotted circle EE′ from the centre O through the point E, and draw

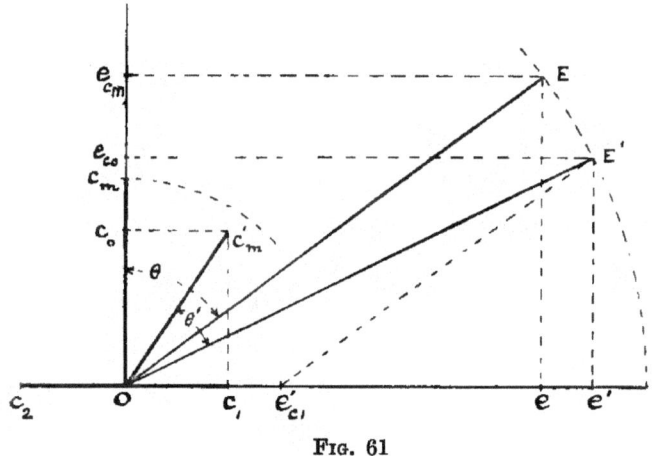

FIG. 61

e'_{c1} E′ parallel to OE. Join OE′ and drop the perpendiculars E′e_{co} and E′e' upon the vertical and horizontal co-ordinates.

The vector OE′ represents the new primary impressed E.M.F. (equal in magnitude to OE). The distance $e'_{c1}e'$ is a measure of the E.M.F. of self-induction due to the magnetising component (Oc_o) of the new primary current, whereas Oe_{co} is the E.M.F. required to force this same component of the total current against the resistance of the coil.

We are therefore in a position to determine the new primary current. The component oc_o is equal to oc_m multiplied by the ratio of $e'_{c1}e'$ to oe, or of oe^{co} to oe_{cm}; and by compounding this with the horizontal component oc_1 we obtain oc'_m, which, it will be seen is in this particular instance *less* than oc_m, the current through the primary when the secondary is on open circuit.

The cause of this is almost identically the same as that which accounts for the behaviour of a choking coil when shunted by a non-inductive resistance (see § 37); and the reduction of the phase difference between the primary impressed E.M.F. and current (see angles θ and θ') which is the necessary result of the current in the secondary coil, by bringing the voltage drop due to ohmic resistance more nearly in phase with the impressed E.M.F., is accountable for this somewhat remarkable result.

INDEX

PRINTED BY
SPOTTISWOODE AND CO., NEW-STREET SQUARE
LONDON